Your Cat is Dead:

Unlocking the Secrets to the Quantum Universe

YOUR CAT IS DEAD:

By Kyle Olechnowicz

Unlocking the Secrets to the Quantum Universe

YOUR CAT IS DEAD:

Unlocking the Secrets to the Quantum Universe

Third Edition, Soft Cover – Published 2012
Printed by Lulu Press in the United States of America
ISBN: 978-0-557-29338-4

This book is dedicated to the one true love of my life, my beautiful wife Natasha. She's been by my side through good times and bad. And she has somehow managed to do the impossible; she's endured the insanity that sometimes comes with being with one whom, speaking politely of, is pretty far from normal. I'll always love you. -Hubs

Acknowledgements

I'd like to first thank my family, starting with my parents, who bore, raised, taught, loved and tolerated me for so long. Without them none of this would be possible. I'd also like to thank my new family, my wife Natasha and four sons: Casey, Zak, Sam, and Parker. You're the force and inspiration that keeps me going every day.

Next, I'd like to thank my BFF's: Sam & Christie; Jon Neihart; my siblings and grandparents; those who inspire me like Albert Einstein, AM talk show hosts and the many others who have shared so much with others, especially the truth.

None of this book would have been possible if not for the groundwork laid by the many great scientists cited in the reference section of this book. They did all the hard stuff. I'd also like to thank all my teachers, especially those who taught me to always ask the question, *why*. And after the first answer, ask *why* again and again, sort of like an annoying four year old.

And of course I need to give it up for our divine Creator, who gave us everything we know and love and much more that we don't, and may never, understand. We owe everything to you.

Table of Contents

List of Illustrations

Introduction to a Quantum

"Reality is merely an illusion, albeit a very persistent one." Albert Einstein

I want to start by explaining a little about the title of this book: <u>Your Cat is Dead</u>. If you actually understand where it came from, you're definitely in the minority. To put your mind at ease, if you're reading this away from home, I don't actually know anything about the health of your cat. Your cat may be just fine. And since cats live an average of about 5,000 days, there's only a small chance that your cat will kick the bucket today. What I do believe, however, is that while you're away from home, your cat is not in a blurry alive-dead state with a probability of each determined by quantum mechanics. And only after an observation is made, the state of the cat changes from the alive-dead state and becomes one of the two. Rather I believe, quite simply, that your cat is always either alive or dead. Here in lies my dilemma. While I trust the results predicted by quantum mechanics, the logic escapes me. And for me, logic is a prerequisite and a requirement for everything.

I know that most introductory chapters or prefaces in books are usually just glanced over, if even read at all. I don't want that to be the case with this book. I think it's crucial that each reader understands some of the challenges we have and are still facing today in the scientific world. I'll discuss some of the challenges in this chapter and later I'll try to offer some explanations.

I also want to point out that very few of the ideas in the book are actually mine. I can take credit for most of the new ones, but people who are far smarter than me laid the groundwork to make my ideas possible. What I've tried to do is to organize and examine what we know, then put the knowledge together to try to solve an elaborate puzzle. In doing so, I've used (and cited) many sources; some of which are not so traditional, like internet articles and online encyclopedias. While I know it is sometimes dangerous to use the internet, it's also often quite helpful. When I used these sources, I did my best to pull out any untruths and only present solid facts. But I'm certainly not perfect and like everything in life, some errors are to be expected.

Unlike some books, this one is intended to be read from start to finish. If you make the mistake of skipping around, you'll not only confuse yourself but you'll make the potentially correct conclusion that I'm completely crazy. It's best to read this book from start to finish and then arrive at your own intelligent well informed conclusion.

I could have spent well over a thousand pages writing this book, but that's not my style. I wanted to present only the most important information in as concise a manner as possible with a little humor thrown in from time to time. Hopefully, this style of writing will keep most of the readers engaged or at least alive.

Over the past few years, I've heard several explanations regarding quantum mechanics. So far, none of these explanations have satisfied my curiosity, some of the ideas have not even been close. Even the best ones have fell a little bit short. Some of the world's brightest people have suggested that quantum mechanics (or quantum physics) is too difficult and beyond the capacity for humans to comprehend. We are therefore told that we should blindly accept the ideas and formulas pertaining to quantum mechanics as facts and not waste our time trying to understand what's happening on the subatomic scale. After all, why should we humans expect objects on such a small scale to obey the same rules that we observe and live by in our everyday macroscopic world?

Many of the quantum rules may seem strange and while they may be different from what we observe in our everyday world, why are we not able to comprehend everything that's happening? The idea that we are not smart enough is very difficult for me and a few other intelligent scientists to accept.

I want to assure you that this book is not some "brief history of the world in a nutshell" or some other type of book that simply summarizes what we already know in science. My intent in writing this book was to present bold new ideas that challenge our way of thinking and to present this information in a way that we can all understand. I want to describe this material in ordinary language that is not at all pedagogical (despite using the word pedagogical in this section). I liken it to painting a beautiful picture that's so complex yet it's impossible to misinterpret because of its naked elegance.

Many people think that subjects such as science, math and physics are very difficult, perhaps even incomprehensible. They compare them to climbing a large mountain or facing some other insurmountable task. It's crucial to find a way to present new and old ideas in a way that we can all understand and appreciate them.

"Knowledge has three degrees -- opinion, science, illumination. The means or instrument of the first is sense; of the second, dialectic; of the third, intuition."
- Plotinus

Many of the greatest scientists, such as Einstein and Feynman, have done exactly this. I'm fully aware that some of the new ideas that I'll present in this book may be completely wrong, but who really cares? I would much rather try and fail then never try at all.

Before I get too deep into this book, I think there's an obvious question that needs to be answered: *What is a quantum?*

After all, I use the word "quantum" in the subtitle of the book and I refer to it a hundred or so times throughout the book so I think it's important that I at least define it. For the purposes, from here on, a quantum can be viewed as a fundamental unit of a quantized physical magnitude. Or more simply put, quantization means that only certain values (of something) are allowed.

Let me give you a very basic example of quantization, one that's completely unrealistic but gets the point across. Let's say that there's a newly invented quantum car that has quantized speeds that allows it to only travel at even whole number speed increments. Again, I know this is not a very realistic but it doesn't matter. Using this scenario, as the car accelerates from 0 to 10 MPH, it will skip over speeds 1, 3, 5, 7 and 9 MPH (and all non-integer values in between). These speed values are considered "forbidden" by the quantum car. At some point, the car will reach its target speed of 10 MPH. Seems kind of crazy, right? Well it should seem strange. In our macroscopic world, the idea of quantized speeds for a car does not make much sense. Nor would any other quantized property on a large scale. But in the atomic world, we observe this type of quantization all the time. Here are a couple of the many examples of quantization:

- electron spin
- electron's orbital angular momentum

While this phenomenon may be extremely difficult to comprehend, the theory of quantized physical properties is currently widely accepted by most experts in the scientific community. It's also, so far, impossible to disprove or discredit. For now, it's the best theory we have and so far has proven to be quite useful.

Goals

The main goal of this book is to offer and explain some of the unknown concepts in science. This is one of many large and very challenging desired outcomes for this book, but hopefully it will be achieved. I'm of the opinion that most of the known concepts in science can be understood by virtually everyone, regardless of their education level. It's important for me to present all the ideas in this book in a simple format. By keeping this book as a thought piece, rather than a mathematical proof, I hope to accomplish this goal.

> "Things should be made as simple as possible, but not any simpler."
> - Albert Einstein

Some of the unknown concepts that I'm going to try to explain include (but are not limited to) dark matter, dark energy, the nature of mass, wave-particle duality, time travel, the Big Bang theory and many more subjects. Again, I want to present these ideas as thought pieces. Someday, people far smarter than me will mathematically or experimentally prove or disprove these new ideas. Albert Einstein once said, "Any intelligent fool can make things bigger, more complex, and more violent. It takes a touch of genius -- and a lot of courage -- to move in the opposite direction." I can assure you that this book will take us in many opposite and strange directions.

Background

I want to spend a very short amount of time discussing some of the history behind and the origins of quantization. I think that it's important and useful to know a little about the history, specifically how we got to where we are today. Only then can we attempt to understand it and begin our search for alternatives. This will only be a brief overview and not at all comprehensive.

Quantum mechanics was first developed in an attempt to provide a better explanation of the detailed workings of the atom. More specifically it was developed to attempt to explain the spectra of light emitted by different atomic particles. With the advent of quantum mechanics, for the first time ever (using the quantum theory of the atom) it was possible to explain a why electrons stay in orbit. This fundamental observation could not be explained by Newton's Laws of Motion or by using Maxwell's Laws of Classical Electromagnetism (both of these laws will be discussed later in this book).

The quantum theory of light, which states that light comes in discrete bundles of energy (called photons), offered the first complete explanation of black-body radiation. A black-body is simply a term that scientist use to describe an object that absorbs all electromagnetic radiation that hits it. In the late nineteenth century, scientists observed that objects at different temperatures emitted different colors, and therefore they emitted different wavelengths of energy. Figure 1 shows that the classical theory predicted that an ideal black body in thermal equilibrium would emit radiation with infinite power, as evident by the slopped line approaching infinity in Figure 1. This led to what became known as the "Ultraviolet Catastrophe."

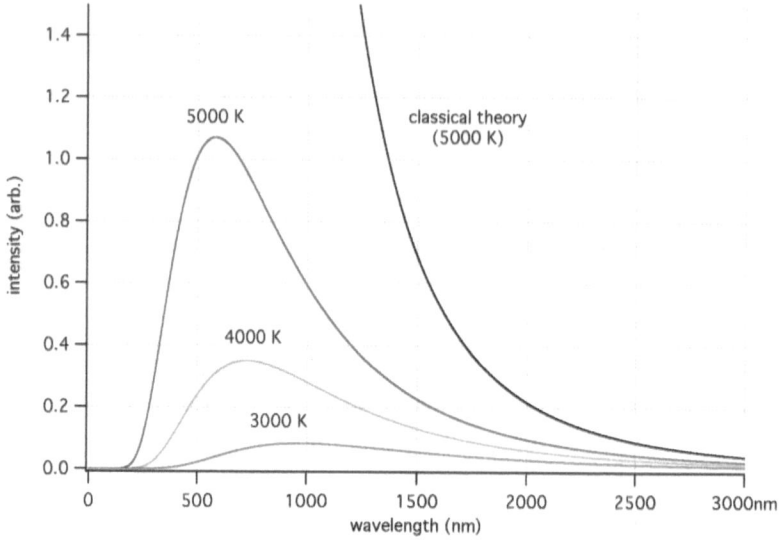

Figure 1: Classical vs. Plank's Black-body Radiation Curves (Wikipedia, 2009)

In the early 1900's, Max Planck found a mathematical formula that fit the experimental data near perfectly. In order to fit the data, however, Planck had to assume that the energy of the light in the cavity was quantized (i.e., an integer multiple of some value). He didn't understand this effect and was quite dissatisfied with his own solution. At the time, his solution was basically just a mathematical fit rather than a comprehensive explanation.

Scientists were later able to make some conclusions by observing the frequency of light of the freed photons during numerous experiments. The scientists were also able to explain why the negatively charged electrons didn't come crashing into the positively charged nucleus. This was the center of much controversy among scientists. Basically, according to classical theory, electron should only stay in orbit for a fraction of a second before they would crash into the nucleus. We know, however, that this isn't true based on our own existence. If it were true, the entire universe, including this relatively unimportant book, would end in seconds.

Scientists explained that when energy is added to an atom, say by heating it up, electrons are able to "jump" to a higher level. When the atom starts cooling down a photon is released, which corresponds to the transition between energy levels (when the atom cools the electron jumps down to a lower level and a photon is freed). In a sense, we can think of this as the conservation of energy. To conserve energy, when an atom cools down, a photon or some other form of energy must be given off for conservation to be fulfilled.

By observing the different frequencies of light, scientists concluded that there were only certain locations that were allowed to be occupied by electrons. The space between these allowed locations were said to be "forbidden". The energy levels, or locations, of electrons became another quantized property in the subatomic world.

Einstein was later able to explain this unusual observation. He went on to propose that the quantization of electromagnetic radiation was the reason for what became known as the photoelectric effect. Many people are surprised to learn that Einstein was awarded the Noble Prize in Physics in 1921 for his work on the photoelectric effect, not for his work on Special or General Relativity, for which he's most famous.

Basically, the photoelectric effect showed the world that higher-frequency light has more energy than lower frequency light. Additionally, it showed that the intensity of light was not proportional to the energy of the light. For me and probably many other people, this is counter-intuitive. For example, if you use a low-frequency light and shine it on a metal, you may not be able to release any electrons, no matter how high the intensity of the light. But if you use a high frequency light, you may be able to knock out some electrons even if the intensity, or brightness of the light is low. Einstein went on to explain that light comes in discreet quantized bundles of energy. These quantized bundles of light energy later became known as photons.

While the idea of quantization is able to offer several explanations to what we observe, to date we have not been able to understand *why* certain properties are quantized. This book was started as a dubious attempt to try to explain some of the strange observations of the quantum world.

Before too much is explained, I need to state the obvious, mainly so you don't think I'm completely crazy (although those of you who know me could make a compelling argument). Most of the ideas that I mention in this piece are going to be controversial. Some, maybe all, will be completely wrong. The ideas challenge what many scientists consider to be modern day scientific facts. These new ideas challenge our elders and go against much of the modern scientific establishment.

However, sometimes in science, like life, we hold dear to ideas and principles for no particular reason. Most of the time when we ask, why, there's a very good reason why. But sometimes, we enslave ourselves to these ideas. Perhaps if we step back and approach a problem from a completely different angle we may see what lies right in front of us: those important things hiding in plain view.

To help put it in perspective, Einstein once said, "We can't solve problems by using the same kind of thinking we used when we created them."

It's also worth mentioning here that much of what we are taught today in our school system is flat out wrong. For some reason, we keep feeding our students the same garbage that's been taught for decades. Let me give you a simple example: ask any child in grade school, "Who discovered that the earth was round?" I would be willing to bet that most of the students would say Christopher Columbus. Why? Because that's what they've been taught for years.

I don't know who the first person was to show or prove that the earth was round, but I'm fairly certain it wasn't Columbus. We know that the ancient Greeks figured this fact out centuries earlier, and possibly others did long before the Greeks.

A nice article, by the Gulf of Maine Aquarium that I read here: http://www.gma.org/space1/nav_map.html on April 21, 2010 described how this simple fact was known centuries before Christ ever walked on earth. In the article, it stated that scholars like Pythagoras in 500 BC based their belief about the shape of Earth on observations about the way the altitudes of stars varied at different places on Earth and how ships appeared on the horizon. As a ship returned to port, first its mast tops, then the sails, and finally its hull gradually came into view. Aristotle, who lived 300 years before Christ, observed that the Earth cast a round shadow on the moon during a lunar eclipse. This would only be possible for a brief moment if the earth was disc shaped (basically when the earth and light from the sun were directly perpendicular to each other). At all other angles of incident, the shadow on the moon would appear oval in shape.

So the bottom line about the information we are told is that just because it's in a book or being taught by teachers in our schools doesn't mean that it's correct. There are so many factors and special interests that influence what we're *permitted* to learn that it would take an entirely new book to even scratch this surface.

Disclaimer

As previously mentioned, some of the ideas in this book will be disproven but perhaps a few will be correct. If only one idea is proven true, or if one idea takes us down a correct path, then a positive contribution to science will have been made. But regardless, by sharing these new ideas there's nothing to lose.

Now, it's important when reading this book, that you note the intentional distinctions between declarations and questions. You can easily spot the differences by noting either a period or a question mark at the end of the sentences, respectively. The questions are solely designed to get you to think, they are not necessarily what I believe. They may, however, be slanted or biased, but please don't think of them as my personal beliefs.

The reason why I chose the title, <u>*Your Cat is Dead*</u> had nothing to do with my initial intentions when I started writing this piece. When I began, I was simply trying to answer the seemingly simple question, "What is Mass?" But this question led to another and that one led to several more. Before I knew it, the simple question about mass became very complicated. I was amazed that I was able to take one of the simplest properties in science and distort it into something that I could barely understand.

If you're like me, you're probably thinking right now, what's so difficult about understanding mass? The total mass of an object is simply the summation of the masses of all the subatomic particles that make up an object, right? It's basically the amount of "stuff" (i.e. electrons, protons, neutrons…) that an object contains. That's one way, perhaps the most common way that we could think of mass.

We could also define mass the way most high schools or college courses in physics do. Normally they teach students that mass is the ability for an object or body to resist acceleration. That's a basic definition that I believe came directly from Newton's 2nd Law: $F=mA$ (The lower case letter denotes a scalar quantity while the uppercase letters denotes a vector quantity. This is a convention that I'll use from time to time throughout this book). This famous equation tells us that the acceleration on an object is directly proportional to the force applied to the object. In this equation, mass (denoted as m) is the constant of proportionality. If the mass of the object is large, then it will require a large force to accelerate it. This is related to Newton's 1st Law: *An object in equilibrium (net force acting on it is equal to zero) will experience no acceleration.* This is sometimes stated as: *Objects in motion remain in constant motion unless acted upon by a non-zero external force.*

These definitions of mass make sense to us because they are what we observe in our everyday lives. Objects with mass do appear to resist acceleration. But is this the complete picture regarding mass? And if so, *why* do objects with mass resist acceleration? And how does the summation of the masses of the subatomic particles cause an object to resist acceleration? These are some very basic questions that started me on my journey and that I'll attempt to answer in this book.

To start this challenging quest, I looked back for some historical precedence. Going back over 2,500 years (about 450 BC), a Greek philosopher, named Democritus, asked some basic questions about cutting cheese into some very small pieces. Basically, he asked a simple question. He asked, "If I am able to cut my feta cheese into smaller and smaller pieces, with my polished bronze knife, will it still be cheese?"

Sounds like an easy question, right? Unknowingly, he was asking a very fundamental question that future scientists would someday answer. We know now that we can keep cutting the cheese until we reach what is considered a single molecule of cheese. If we keep cutting the molecule, we can get the cheese down to atoms and fundamental particles (electrons, protons, neutrons…). I think it is a safe bet that Democritus was unaware he was on the verge of stumbling upon the idea of atoms and other subatomic particles. But his questions did lay the groundwork for much of modern day atomic theory.

Using the ideas from Democritus, the Greeks coined the word átomos which means something that cannot be divided further. We now know that using this as a definition of the atom would not be completely accurate because the atom can be divided further (into neutrons, proton, electron, quarks…). I only mention this because many people believe that mass comes from the addition of all the tiny subatomic particles that make up a substance. So if I connect this basic idea to our previous high school definition of mass, we could infer a more complete definition: *Mass is an object's ability to resist acceleration caused by the summation of the subatomic particles in that object.*

This idea about mass may seem trivial, but it's not. I especially don't want to understate the importance of mass as it relates to the atom and to science. In fact, in *"The Feynman Lectures on Physics"*, by Dr. Richard Feynman, he stated:

"If, in some cataclysm, all scientific knowledge were to be destroyed, and only one sentence passed on to the next generation of creatures, what statement would contain the most information in the fewest words? I believe it is the atomic hypothesis (or atomic fact, or whatever you wish to call it) that all things are made of atoms - little particles that move around in perpetual motion, attracting each other when they are a little distance apart, but repelling upon being squeezed into one another. In that one sentence you will see an enormous amount of information about the world, if just a little imagination and thinking are applied."

For Feynman to consider this the most informative statement humanly possible to pass down to future generations shows us how significant he thought of the atomic world. I think it's just as important for us to understand the complexities of relating the atomic world to mass. But making this connection may not be as simple as we think.

We'll examine mass and the atom in much more detail throughout this entire book. To conclude this introductory chapter, I'd like you to consider this basic thought problem. Much of the background needed for this problem will be explained later. But it should still make sense to everyone.

Let's say, we are able to completely isolate a sub-atomic particle like a proton. *The Standard Model of Physics*, which we'll explore in Chapter 3, states that a proton is composed of two up quarks and one down quark. For now, let's not worry about what that means and let's assume that the Standard Model is correct. Let's also assume that we can isolate one of the quarks (we're assuming this is possible, despite what the *Heisenberg's Uncertainty Principle* states). We are also going against the widely held belief that we can never pull out a single quark from inside a particle, such as a proton. Basically, the idea is that even if we could provide enough energy to overcome the binding energy of the quarks, all we would do is create a quark-anti-quark pair. This is a process known as "pair creation" during which the new quark replace the one that was removed and the anti-quark combines with the removed quark to form a meson. We'll come back and discuss this in greater detail in Chapter 3.

But this is just a thought exercise, so let's say it is possible to pull out and completely isolate a quark. Now, assuming all the preceding to be correct, is it possible to divide the quark even further? If not, what would prevent us from doing so (besides our lack of technology)?

More simply put, can we can just keep cutting and cutting forever? We certainly don't know if the quark is the final composite point. It might be or it might not be. It's not really that important if the quark is the final point; the intriguing question at hand is whether we can keep cutting and cutting forever. This basic idea brought me to a key assumption that I thought was necessary.

Assumption: There must be a point where we cannot divide particles any further.

This assumption is based on speculation and a somewhat educated guess. I think that it's important and necessary to start with an assumption regarding this, if we're to truly understand mass. The assumption may turn out to be incorrect, but I'll use this as a building block, and expand on it or crash it down as we move forward.

This assumption seems basic and safe, but it's not. If this assumption is true, then how can an object, occupying a finite space, be indivisible? This is only possible if there's an error with the assumption or a mistake in our current perception about what constitutes matter and space; my vote is for the latter. Either way, there's much more to mass than first meets the eye.

Rest of the Book

Congratulations, you just finished the first chapter which was a brief introduction to a quantum and to quantum mechanics. The next few chapters (Chapters 2-4) focus on developing the background to start exploring some new ideas in physics. In the next 3 Chapters, we will briefly explore Newtonian Mechanics (as well as the science leading up to this point in time), quantum physics, modern physics and all the groundwork up to and including the work of Albert Einstein (not necessarily in this order).

The remaining chapters discuss the origins of mass, space and time and paint a qualitative picture of our universe. They also discuss how to view and understand the universe around us and in the chapters, we will go so far as to present a unique unified theory in a qualitative manner that is understandable by even grade school students. I agree with a not-so-famous quote by Einstein when he said something of the sort, "You do not truly understand a topic until you can explain it to your grandmother." This explanation of a unified theory should be comprehendible by all while the mathematics may not be as easy to digest. The conclusion of the book will focus on our way forward and how to use this new knowledge.

History & Basics of Quantum Mechanics

"Anyone who is not shocked by quantum theory has not understood it." Niels Bohr

I've never been a huge fan of quantum mechanics, actually I've never even been a little fan. In graduate school, I found quantum mechanics to be by far my most difficult subject, in fact I found it nearly impossible to comprehend. Despite my lack of comprehension, I somehow made it through this class. In most of the quantum mechanics lectures, students were asked to push the invisible *"I Believe"* button. This means they were asked to just accept what was being taught without first understanding the material. Personally, I think my memory is relatively poor, so for me it's very important to understand or to be able to derive what's happening. I don't like, nor have I ever liked, trusting everything I've been told or everything that I've read without understanding it. As such, when I discuss quantum mechanics throughout this book, I'm going to keep it as a qualitative discussion trying first to grasp the information; unfortunately however, some formulas and descriptions will need to be mentioned and explained.

It was shortly after I started contemplating about mass that I realized the importance of space in describing it. I'm not talking about outer space, just the space around us that we observe and interact with in our everyday lives. You'll find that I tend to use the term space fairly loosely in this book. During my brief tenure on earth I've gained a great respect for space, so much so that I've dedicated an entire chapter to it. When I talk about space in this book, you can assume that time in some form is attached to it. We'll explore this idea much deeper in later chapters. For now try to imagine this very basic mental illustration to help understand space. I'll refer back to this scenario later on in the book:

On your way to work, you look across a fairly empty field. At the opposite end of the field there's a tree. Let's say that the field is completely flat and uniform throughout.

Q: How many possible ways could you walk across the field to reach the tree?

A: There's <u>seemingly</u> an infinite number of ways that one could reach the tree.

That may seem like a pointless example right now, but are objects, like atoms or electrons, in open space any different? For example, if we shot an atom at a screen, I suspect that most people would assume the atom would travel in a straight line. Maybe some of the more educated people would think that the atom would be slightly deflected by gravity or by another force. Personally, I wouldn't expect the atom to zigzag all over the place and reach the screen at some random location. But much of quantum mechanics tells us just that. Simply put, quantum mechanics is a tool that allows us to calculate probabilities of future events on the subatomic scale. These probabilities are very useful in subatomic calculations. But these quantum calculations do not explain *what's* happening or *why* it's happening. That's where I have a problem. That's what I want to try to understand.

Schrödinger's Equation

I stated before that it would be unfortunately necessary to have a small amount of math and equations in this book. As such, now it's time for our first challenging formula. We'll start with one of the most dreaded formulas of all: At the heart of quantum mechanics is the formula known as the *Schrödinger Equation.*

The time dependent form of Schrödinger's Equation is:

$$-\frac{\hbar^2}{2m}\frac{\partial^2 \psi}{\partial x^2} + V(x)\psi = i\hbar\frac{\partial \psi}{\partial t}$$

Please don't look at this formula like most people do and panic. It's not time to jump off a bridge and the formula is not too difficult to understand. Simply put, the right side of the Schrödinger equation describes how the wave-function (ψ) changes with time and the left side of the equation is related to a particle's energy.

m is the mass of the particle moving in the one-dimensional potential $V(x)$

The one-dimensional potential $V(x)$ describes the forces acting on the particle.

$\dfrac{\partial^2 \psi}{\partial x^2}$ describes how the wave-function (ψ) changes from one place to another.

Using this equation, Schrödinger computed the spectral lines for hydrogen by treating a hydrogen atom's single negatively charged electron as a wave, $\Psi(x, t)$, moving in a potential well, V, created by the positively charged proton. This computation reproduced the energy levels of the Bohr model (Wikipedia, 2009).

Schrödinger's equation describes how a wave and its properties evolve over time. It can be used to solve many types of subatomic problems. Luckily for you, none of these problems will be done in this book. Schrödinger's equation is an equation that's discussed about all the time, so it's definitely worth at least mentioning. We'll also come back to it and examine it in more detail later on in this book.

Prior to adopting Schrödinger's equation, scientists used matrix mathematics to describe the evolution of a wave. In matrix mathematics, A x B (that's A times B not A cross B) does not always equal B x A. Nowadays, nearly all scientists have learned and use Schrödinger's equation due to its relative simplicity over the matrix method.

One of the most unappealing aspects of quantum mechanics for me is that quantum mechanics was founded based on empirical observations from experiments rather than from being derived by other fundamental laws. This was also one of the sticky points that also made the theory so unappealing to Albert Einstein. An additional drawback to quantum mechanics is that sometimes when we apply it to a system we get some very surprising results that cannot be explained or understood. For example, quantum mechanics tells us that it's possible for particles to disappear and then reappear at some new random location (at least seemingly random). Quantum mechanics also tells us that it's possible for particles to be in two places at the same time. We'll discuss these unusual results in greater detail when we discuss wave-particle duality and other aspects of quantum mechanics. But for now, it's worth noting that there are many surprising results. Amazingly, however, to my knowledge no experiment has ever been performed that has been able to disprove quantum mechanics. The theory appears to hold true in all experiments.

I'd like to reiterate that Albert Einstein was not a huge proponent of quantum mechanics, even though many scientists consider him to be the founding father. One of his most

"Science without religion is lame, religion without science is blind."

- Albert Einstein

famous quotes was, "[T]hat He would choose to play dice with the world," he wrote of God. Most references to this quote rephrase this as, "God does not play dice with the universe." Randomness and uncertainty were not the only reasons why Einstein rejected quantum mechanics. There are several more reasons that we'll explore as we go on.

As mentioned, I believe that a significant drawback for Einstein was the uncertainty with quantum mechanics; specifically the inability to predict the exact properties (location, speed, energy, momentum…) of subatomic particles. Einstein believed in an orderly, deterministic nature of the universe; that is, everything in the universe can be determined. A minority of other scientists also have trouble accepting that the best we can know is simply a probability. To help understand this, let's consider the following example:

Let's assume we have an unlimited knowledge of all the forces of nature and we also have infinite computing resources. Would it then be possible to predict exact values of the subatomic world? For example: Could we know the exact location of a certain atom one second from now? How about two minutes from now? Quantum mechanics tells us, No.

In the macroscopic world, we can use basic physics to predict many things. Let's consider a game of pool, for example. When a player hits the cue ball to try to sink another ball in a

pocket we can make a prediction, with a given margin of error, as to whether he will hit the target ball in the pocket. The more information we know, the smaller the margin of error will be. To make a very accurate prediction, several variables would need to be known to a high degree of accuracy (i.e. speed, spin, slope of table, angle of impact, bumps on the table, temperature...). The point is that we can grasp and understand what's happening on this macroscopic level. So why do we have so much trouble in the subatomic world?

If we could make these same types of predictions (with a high level of accuracy) on the subatomic level, could we then predict everything? If we could predict the exact future behaviors of atoms, could we perhaps even be able to predict how people would behave? We people are, after all, simply compositions of subatomic particles that are set into motion according to the laws of the universe. One could argue that depending on how the universe behaves and the chemical balances in our brains (which are caused by actions of subatomic particles) that we may be "forced" to make certain decisions. However, I don't think it's quite as simple as this. In fact, this idea would also call into question the reality of free will. But could it be possible that we don't actually have free will? Is it possible that we have no choices in our decisions? That may be too philosophical of a topic to get into here, but you get my point. Personally, I think I do have free will. And I don't think a judge or a spouse would accept "the laws of the universe" as a valid excuse for lapses in judgment. But good luck if you choose to use that defense.

Chaos

For a moment, I want to go against the grain of quantum mechanics and assume that the atomic world is deterministic, similar to what Einstein believed. What effects, if any, would this bold assumption have on science? One consequence of having an infinite knowledge about forces and infinite computing resources is that we may even disprove the notion of chaos. However, I don't think that's probable and here's why:

The name chaos comes from the fact that a system's behavior is apparently disordered. Chaos theory, however, is really about finding the underlying order in apparently random data. Chaos was first discovered and described by Edward Lorenz in 1960. At the time, Lorenz was working on predicting the weather. He had a computer set up with twelve equations to model the weather. This computer program was capable of theoretically predicting what the weather might be at a later time in the future.

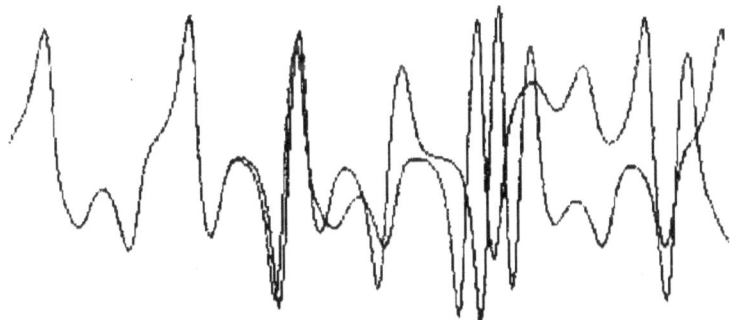

Figure 2: Lorenz's experiment - difference between the starting values is .000127 (Stewart, 2002)

After the experiment was started, at some later time, Lorenz wanted to run the experiment on a particular sequence again. However, to save time, he started in the middle of the sequence, instead of the beginning (shortcuts in science are like going from nowhere to nowhere even faster). He entered the number off his printout and left the room while he let the experiment run. Computer speeds were not like they are today and it took some time to run the sequence.

When Lorenz came back about an hour later, the sequence had evolved quite differently than it had before. Instead of the same pattern he had seen before, it diverged from the pattern, ending up wildly different than the original results (Figure 2 shows how these two sequences evolved). Eventually, Lorenz was able to determine what went wrong with his experiment. The problem was that the computer he used stored the numbers to six decimal places in its memory. To save paper, he only had it print out three decimal places. In the original sequence, the number was .506127. In the new sequence, he had only typed the first three digits, that is, .506.

By all conventional ideas of the time, the experiment should still have worked. At a minimum, Lorenz should have found a sequence very close to the original sequence. After all, most experiments are successful if the results have accuracy within three decimal places (My standards are a bit lower. I consider my experiments to be successful if nobody loses any appendages). At the time, it was believed that the fourth and fifth decimal places couldn't have a huge effect on the outcome of the experiment. Lorenz proved that this idea was not always correct.

This effect later became to be known as the "butterfly effect". Basically, the amount of difference in the starting points of the two curves is so small that it is comparable to a butterfly flapping its wings.

"The flapping of a single butterfly's wing today produces a tiny change in the state of the atmosphere. Over a period of time, what the atmosphere actually does diverges from what it would have done. So, in a month's time, a tornado that would have devastated the Indonesian coast doesn't happen. Or maybe one that wasn't going to happen does." (Stewart, 2002)

This phenomenon is also sometimes known as "sensitive dependence on initial conditions." This means that just a small change in the initial conditions can drastically change the long-term behavior of a system. Such a small amount of difference in a measurement might be considered experimental noise or an inaccuracy of the equipment for most experiments. These small deviations are impossible to avoid in even the most isolated and advanced lab. With a starting number of 3 for an experiment, the final result can be entirely different from the same system with a starting value of 3.000000001. It is nearly impossible to achieve this level of accuracy in any controlled environment. Good luck trying to measure something to the nearest billionth of an inch. (IHMO.com. 2010)

At this point in the book, it's worth noting that not all experimental systems evolve into a chaotic system. In fact, most do not. Thankfully, many systems are not drastically dependent on the initial conditions. But when they are, chaos needs to be accounted for and very careful measurements must be made.

That was a quick and very brief description of chaos theory but it may be useful for the information that follows. Then again, it may be completely worthless. One of the main goals of this book is to understand if systems that we view as chaotic could be viewed as being deterministic, so I believe that chaos theory will be helpful.

"Chaos is the score upon which reality is written."
-Miller, H. Valentine
Tropic of Cancer.

I'd like to give you a simple example to show how chaos may be relevant in this book:

In many regards we can consider the evolution of the universe as a chaotic system. You see, during the beginning moments of our universe, any slight deviation in speeds, accelerations, or the collisions of the particles could have created a universe that is far different from the universe that we see today. Therefore, the evolution of the universe was, and still is, highly dependent on the initial conditions.

But if we could better model and understand the laws of the universe, would it be possible to accurately determine the exact evolution? *Relating this to quantum mechanics, could it be possible that the probabilistic nature of events on the subatomic scale may simply be the perception we have due to our own scientific limitations and the lack of understanding about the universe around us?* This is the ultimate question to answer and one of my personal goals.

We'll come back and explore these and many other theories and properties of quantum mechanics later in this book. This chapter was written to simply give you a general overview of the basics and history of quantum mechanics and chaos. There's much more to follow, including many new and exciting ideas.

Where We Are Today

"The important thing is not to stop questioning. Curiosity has its own reason for existing." Albert Einstein

I want to spend a little time discussing what scientists know, or at least think they know at the present time. I'll also discuss some of the current theories that help to explain the universe as we currently see it. Many of these ideas and theories are less than a few decades old, some are centuries old, and some are brand new. Obviously, all the current theories in physics are not presented in this chapter. This would take several volumes of very large books to put together. For our purpose, I've only included theories that will provide useful information later. This section is a bit long but all the information will be necessary. Many of these ideas will lay the groundwork for some completely new and groundbreaking theories.

MV or MV²?

This is a good starting point for our discussion and this concept will prove to be invaluable to Einstein's work a century and a half later. In the late 18th century, both Leibniz and Newton were trying to define what quantity of motion was conserved. Newton with mv, and Leibniz used mv² (the ½ was added later).

Leibniz contrasted Descartes' quantity of motion with his "living force". As Leibniz stated in his <u>Specimen Dynamicum</u>: *"I concluded that besides purely mathematical principles subject to the imagination, there must be admitted certain metaphysical principles perceptible only by the mind, and that a certain higher, and so to speak, formal principle must be added to that of material mass, since all the truths about corporeal things cannot be derived from logical and geometrical axioms alone, namely, those of great and small, whole and part, figure and situation—but that there must be added those of cause and effect, action and passion, in order to give a reasonable account of the order of things."*

If you've done a well crafted physical experiment involving energy, you may have found what Leibniz determined: that a moving object's ability to effect change is determined not by *mv*, but by a "higher notion" outside the realm of our sense perceptions, proportional to the mass the square of the velocity, or mv^2.

Leibniz continued in his <u>Preliminary Specimen</u> (1691): *"When I discovered these things, I judged that it was worth the trouble to muster the force of my reasonings through demonstrations of the greatest evidence, so that, little by little, I might lay the foundations for the true elements of the new science of power and action, which one might call dynamics."*

Leibniz' predictions went widely unknown for some time, so we can't finish this very brief section without mentioning one of the greatest female scientists of all time, Emilie du Châtelet. Emilie had no formal scientific training but she understood the value of mathematics in science. As an Aristocrat, she acquired tutors and delved into the more technical scientific literature. She also studied the more philosophical writings of Leibniz, which had been dismissed by her peers.

She was interested by Leibniz's arguments, but found little real proof. She tried to flesh out his ideas by performing her own experiments, but without success. Eventually, she uncovered experimental work by Willem Gravesande of the Netherlands, who had devised a very ingenious yet simple demonstration. He dropped weights into a block of clay and determined the effect of the speed of the weights on the depth they sunk into the clay. In performing this experiment, a simple relationship was found: the depth was proportional to the square of the speed, just as Leibniz had predicted.

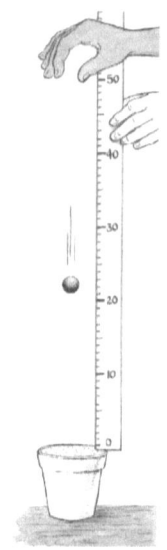

Emilie Du Châtelet was then able to combine the experiment with the theory, and she went on to publish her ideas in the form of a textbook for children. This was the only avenue open to her, as she was not recognized by the academe. Unfortunately, she didn't live long past this publication, dying in her early 40's from complications due to childbirth.

Her work showing that a moving object's ability to effect change is proportional to the mass times the square of the velocity, or mv^2 would be critical for Einstein's future work as we will show later in this book.

Newton's Laws

We'll go back and forth in time through this book, but for now we need to keep building our way to the present. We'll continue by exploring the key ideas of perhaps the greatest scientist of all time, Sir Isaac Newton. Prior to Newton, there were some very important and complicated theories and ideas that came to light as we've already discussed (many were omitted to save space) but after Newton, nothing in the scientific community would ever be the same.

Let's begin our discussion about how Newton changed our understanding of the universe by enumerating his important three Laws of Motion. These are very famous and widely accepted laws. They are basically stated as:

I. Every object in a state of uniform motion tends to remain in that state of motion unless an external force is applied to it.

We can recognize this first law as essentially Galileo's concept or vision of inertia. This law is often referred to as simply the "Law of Inertia".

II. The relationship between an object's mass m, its acceleration A, and the applied force F is $F = mA$. Acceleration and force are vectors (as indicated by their symbols being displayed in slant, bold, uppercase font). In this law, the direction of the force vector is the same as the direction of the acceleration vector.

This is arguably the most powerful of Newton's three laws, because it allows quantitative calculations of

Apollo 8 Astronaut, Bill Anders: "I think Isaac Newton is doing most of the driving now."

dynamics: *How do certain velocities change when forces are applied?*

It's worth noting that there is a fundamental difference between Newton's 2nd law and the dynamic predictions of Aristotle. According to Newton, a force causes a change in velocity (which we refer to as acceleration). The object does not maintain a constant velocity as Aristotle proposed.

This is sometimes summarized by saying that under Newton's law, $F = mA$, but under Aristotle's law, $F = mV$

where V is the velocity

Thus, according to Aristotle there is only a velocity if there is a force, but according to Newton an object with a certain velocity maintains that velocity unless a force acts on it to cause acceleration (that is, a change in the velocity). This was a major paradigm shift in science and an extraordinary insight, not at all intuitive (by Newton).

Aristotle's view seems to be more in line with common sense, but that's because of a failure to appreciate the role played by frictional forces. Once we take into account all of the forces acting in a given situation, it's the dynamics of Galileo and Newton, not of Aristotle, which are found to be in agreement with the observations.

III. For every action there is an equal and opposite reaction.

This law is exemplified by what happens if we step off a boat onto the bank of a lake: as we move off the boat in the direction of the shore, the boat tends to move in the opposite direction, potentially leaving us face down in the water, wishing we understood Newton's 3rd law.

Conservation of Energy

The Law of Conservation of Energy is an empirical law of physics. It states that the total amount of energy in an isolated system remains constant over time. The total energy is said to be conserved over time. A consequence of this law is that energy can neither be created nor destroyed; it can only be transformed from one form to another. The only thing that can happen to energy in a closed system is that it can change form. For instance, potential energy can become kinetic energy, like when a cat falls out of a high tree and goes splat.

Albert Einstein's famous Theory of Relativity showed the world that energy and mass are the same thing (we'll explore this in great detail later). Thus in closed systems, both mass and energy are conserved separately, just as was understood in pre-relativistic physics. The new feature of relativistic physics is that "matter" particles (such as those which make up an atom) could be converted to non-matter forms of energy, such as light; or kinetic and potential energy (sometimes seen as heat). However, this conversion does not affect the total mass of systems, since these forms of non-matter energy still retain their mass through such conversion.

Conservation of Mass

Antoine-Laurent Lavoisier (1743-1794) first showed that all the seemingly diverse bits of tree, rock and iron on Earth (all the "mass" there is) really were parts of a single connected whole.

Through decades of meticulous experiments, aided by his able wife Marie Anne, Lavoisier proved that the substances that fill our universe can be burned, squeezed, shredded, or hammered to bits, but they won't disappear. The different sorts floating around just combine or recombine. The total amount of mass, however, remains the same. This principle, *The Law of Conservation of Mass*, was one of the great scientific achievements of the 1700s.

Lavoisier explored, for example, what happens when objects rust. While intuition might suggest that a rusted piece of metal weighs less than a pristine one, Lavoisier took nothing on trust. He built an entirely closed apparatus and set it up in his house. He put various substances in the apparatus, sealed it tight, and applied heat or started an actual burn to speed up the rusting. Once everything had cooled down, they took out the mangled or rusty or otherwise burned-up metal and weighed it, and also carefully measured how much air was lost. Each time they got the same and unexpected result. What they found, in modern terms, was that a rusted sample does not weigh less. Nor does it weigh the same. It weighs more.

What was happening to make this sample weigh more? There was the same amount of stuff overall, yet now the oxygen (that had been in the gases floating above) was no longer in the air. But it had not disappeared. It had simply stuck onto the metal. With his state-of-the-art weighing machine, Lavoisier showed that matter can move around from one form to another, yet it will not burst in and out of existence. This was a remarkable discovery and is still one of the cornerstones of science.

Faraday's Laws

Michael Faraday was an English chemist and physicist who

contributed to the fields of electromagnetism and electrochemistry.

Although Faraday received little formal education and knew little of higher mathematics such as calculus, he was one of the most influential scientists in history; many regard him as the best experimentalist in the history of science.

His research regarding the magnetic field around a conductor carrying a DC electric current allowed Faraday to establish the basis for the concept of the electromagnetic field in physics, which was subsequently enlarged upon by James Clerk Maxwell (as we will discuss later). Faraday established that magnetism could affect rays of light and that there was an underlying relationship between the two phenomena. He similarly discovered the principle of electromagnetic induction, diamagnetism, and the laws of electrolysis.

His inventions of electromagnetic rotary devices formed the foundation of electric motor technology, and it was largely due to his efforts that electricity became viable for use in technology. It's still used in some form by most people in everyday life.

As a chemist, Faraday discovered benzene and invented an early form of the Bunsen burner and the system of oxidation numbers, and terminology such as anode, cathode, electrode, and ion. Faraday ultimately became the first and foremost Professor of Chemistry at the Royal Institution of Great Britain, a life-time position.

Faraday was an excellent experimentalist who conveyed his ideas in clear and simple language; his mathematical abilities, however, did not extend as far as trigonometry or any but the simplest algebra. It was James Clerk Maxwell who took the work of

Faraday, and others, and consolidated them with a set of equations that lay at the base of all modern theories regarding electromagnetic phenomena. On Faraday's uses of the lines of force, Maxwell wrote that they show Faraday *"to have been in reality a mathematician of a very high order – one from whom the mathematicians of the future may derive valuable and fertile methods."*

The SI unit of capacitance, the farad, is named in his honor.

Maxwell's Equations

This section will conceptually describe each of the four Maxwell equations, and it will also explain how they link together to describe the origin of electromagnetic radiation such as light. I think this type of explanation is more useful than a complex presentation and derivation, followed by a summary of the formulas. The formulas can be quite intimidating if presented alone. And by understanding the laws, one can derive the formulas or at least come fairly close.

Side note: It's worth noting here that Maxwell owes a great deal of credit for his work to his elder and friend, Faraday. Faraday truly laid the groundwork for Maxwell's equations. Faraday was thanked and elated when Maxwell presented his mathematical solutions to many of Faraday's ideas near the end of Faraday's life.

- *Gauss' law* relates the electric charge contained within a closed surface (known as a Gaussian surface) to the surrounding electric field. It describes with mathematical clarity how the divergence of an electrical field is affected by charges (electric field lines diverge from positive charges and are drawn towards negative charges). It also states that the total electric flux through a Gaussian surface is unrelated to the shape and size of that surface.

- *Gauss' law for magnetism* states that the total magnetic flux through a Gaussian surface is zero. This is due to the fact that real world magnetic charges comes in pairs (referred to as dipoles), with the two charges giving rise to opposite magnetic field divergences which cancel each other out. The theoretical single magnetic charge is referred to as a magnetic monopole. Magnetic monopoles have never been observed, but if they do exist, this law would need to be modified. I've been searching for magnetic monopoles for some time and have so far been unsuccessful.

- The Maxwell equation that is known as *Faraday's law* was named by Oliver Heaviside. It describes how a changing magnetic field is related to the induced electric field. This aspect of electromagnetic induction is the operating principle behind many electric generators. It should be noted, however, that this particular equation only caters for the time varying aspect of electromagnetic induction, and does not apply to the motion induced aspect, and 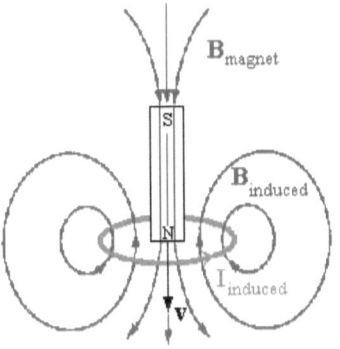 that it takes on a different mathematical form than Michael Faraday's original law. In the original Faraday's law of induction, both aspects of electromagnetic induction are catered for.

- *Ampère's law with Maxwell's correction* states that magnetic fields can be generated in two ways: by electrical current (this was the original Ampère's law) and by changing electric fields. The idea that a magnetic field can be induced by a changing electric field follows from the modern concept of displacement current which was introduced to maintain the solenoidal nature of Ampère's law in a vacuum capacitor circuit. This modern displacement current concept has the same mathematical form 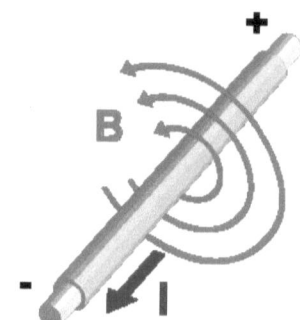 as Maxwell's original displacement current. Maxwell's current applies to the polarization current in a dielectric medium, and it sits adjacent to the modern displacement current in Ampère's law.

Maxwell's correction to Ampère's law was a particularly important advancement. In 1864, Maxwell derived the electromagnetic wave equation by linking the displacement current to the time-varying electric field that's associated with electromagnetic induction. This is described in Maxwell's *A Dynamical Theory of the Electromagnetic Field*, where he wrote:

"The agreement of the results seems to show that light and magnetism are affections of the same substance, and that light is an electromagnetic disturbance propagated through the field according to electromagnetic laws."

The extension to displacement current applies in the pure vacuum. This has been interpreted by some to mean that a changing electric field can produce a magnetic field, and vice-versa. Under this interpretation it follows that even with no electric charges or currents present, it is possible to have stable, self-perpetuating waves of oscillating electric and magnetic fields, with each field driving the other. The physical parameters of transverse elasticity and density, which Maxwell used to calculate the speed of these electromagnetic waves, have been replaced by two easily-measurable physical constants, the electric constant and the magnetic constant.

The speed calculated for electromagnetic radiation exactly matches the speed of light. And it should because light is one form of electromagnetic radiation (as are X-rays, radio waves, and others). Maxwell thus unified the previously separate fields of electromagnetism and optics.

$E=mc^2$ (Round 1)

There's some debate over how Einstein came up with his famous equation, $E=mc^2$. But the important point for this book is that he DID. Nonetheless, we'll discuss this briefly.

Over time, physicists became use to multiplying an object's mass by the square of its velocity (i.e. mv^2) to come up with a useful indicator of the object's energy, as we discussed with the work of Leibniz' and Emilie du Châtele. If the velocity of a ball or rock was 10 mph, then they knew that the energy it carried would be proportional to its mass times 10 squared. If the velocity is raised as high as it could go, to 671 million mph (the speed of light), the energy an object will contain should be revealed when you look at its mass times c squared, or its mc^2. But this is certainly not a proof.

So how did Einstein finally solve this puzzle? After several years of thinking, he finally had an insight that changed the course of the world. So what was this insight?

All Einstein did was to assume that the speed of light was constant. He assumed that nothing could go faster than the speed of light, and that all light traveled at the same basic speed, regardless of the observer.

Up to this time, it had not been established that the speed of light was constant. Everyone thought that time and distance were constants, but that the speed of light, like the speed of everything else in the universe, was variable. But Einstein was willing to consider that what everyone believed about light, time, and distance might actually be wrong.

So he took this assumption, that the speed of light was a constant, and he returned to the mathematical and electromagnetic equations that were worked out years before, most by Maxwell. He then plugged in the letter "c" (a constant) to represent the fixed speed of light (whatever it might be) and he arrived at $\rightarrow e=mc^2$

If the speed of light was truly a constant, as he predicted, then that energy and matter must be one and the same (energy equals matter times the speed of light squared). Not only must energy and matter be the same, but the amount of energy in even the tiniest piece of matter, like the head of a pencil, is phenomenal, far exceeding any conventional bomb or explosive. The mass/energy equivalent of an entire pencil is about the same as an atomic bomb.

Also, if the speed of light were constant, Einstein also reasoned that time and distance must therefore be relative. But this was totally contrary to what everyone, including the world's leading scientists, believed.

This didn't stop Einstein, however. In 1905, he published his argument, including his conclusion that e=mc^2, in a three-page paper entitled *"Does the Inertia of A Body Depend on Its Energy Content?"* The paper had no footnotes and not one single reference to support it.

Now, you may ask, is it possible to derive his famous equation using Newtonian mechanics? Let's give it a try:

Starting from the Kinetic energy equation, as we discussed earlier: $KE = \frac{1}{2} mv^2$

Then we substitute the Newtonian equation of motion (for constant acceleration):

$$v^2 = u^2 + 2as$$

this gives:

$$KE = \frac{1}{2} m(u^2 + 2as) = \frac{1}{2} m u^2 + mas$$

(Where: *v = final velocity, u = initial velocity, a = acceleration, s = displacement*)

Now we have the initial Kinetic Energy $= \frac{1}{2} mu^2$ and

Potential energy $= mas$

So, now the equation instead of KE, we can call total energy:

Total energy = initial KE + potential energy $= \frac{1}{2} m u^2 + mas$

We now want to consider the case when $\frac{1}{2} m u^2 = mas$ so we have:

Total energy $= 2 (\frac{1}{2} m u^2) = mu^2$

And when u =c (such as for light in vacuum free of influences) this is

$$E=mc^2$$

There's several other ways to arrive at this same result, this was just one example.

At this point in the book it would be great to start a discussion on Einstein's special and general theories of relativity. But due to their importance, as they relate to mass and to the rest of the content of this book, I've decided to dedicate the next chapter entirely to these theories. I apologize for any anxiety this may cause.

Fundamental Forces

All the interactions in the universe are believed to be governed by four fundamental forces: strong, weak, electromagnetic and gravitational. There is little evidence to date that contradicts this theory.

Physicists, like me, are actively trying to find one theory that would describe all the forces in nature as a single unified law. This was the subject (some would argue more of an obsession) of Einstein's research for close to the last 20 years of his life. So far, scientists have succeeded in producing a single theory that combines the weak and electromagnetic forces (called electroweak force). The strong and gravitational forces have not been so easy to combine so they are not yet described a combined theory.

Below, the four fundamental forces are described and the different properties of each are outlined. This is the most basic description of these forces, while still being accurate, that I could find. (Wikipedia. 2010)

- The *strong interaction* is strong, as noted by its name, but it is very short-ranged. It acts only over ranges of order 10^{-13} cm and is believed to be responsible for holding the nuclei of atoms together. It is 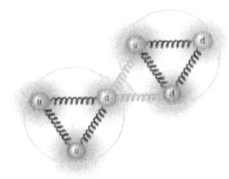 normally attractive, but can be repulsive in some circumstances.

- The *electromagnetic force* causes the electric and magnetic effects such as the repulsion between like electrical charges. It is long-ranged, but much weaker than the strong force. It can be attractive or repulsive,  and it is believed that it acts only between pieces of matter carrying an electrical charge.

- The *weak force* is responsible for radioactive decay, neutrino interactions and it initiates the process known as hydrogen fusion in stars. Weak interactions are most noticeable when particles undergo beta decay and in the production of deuterium  and then helium from hydrogen that powers the sun's nuclear process. It has a very short range and as its name states, it is weak.

- The *gravitational force* is also weak, but it has a long range. Additionally, it is believed to be always attractive, and it acts between any two pieces of matter in the universe since mass is its source. It's also what makes cats  fall from the trees and keeps dead birds from littering the skies, post mortem.

Below is a fairly concise and helpful table showing some of the properties of these four fundamental forces. We'll explain the column in the table labeled "Mediating Particle" in the next section which relates to the standard model. For now, it's useful information that should help us gain a sense of the range and the relative strengths of the various particles.

FUNDAMENTAL FORCES			
Interaction	**Relative Strength**	**Range**	**Mediating Particle**
Strong	1	Short	Gluon
Electromagnetic	0.0073	Long	Photon
Weak	10^{-9}	Very Short	W,Z
Gravitational	10^{-38}	Long	Graviton

Table courtesy of University of Guelph, Guelph, Ontario (Canada)

The Standard Model

We can't address particle physics, which is one of our long term goals, without discussing the so called standard model. Hopefully, I'll be able to describe the standard model in a way to keep it from being unnecessarily complicated. It's an extremely important model and we'll refer back to it throughout the remainder of the book so we need to at least understand the basics.

The standard model is one of the most widely accepted explanations (and one of the most complete) of the subatomic world in modern physics. The standard model of particle physics is able to describe three of the four known fundamental interactions and the elementary particles that take part in these interactions. These particles are believed to make up all visible matter in the universe. Some scientists have suggested that some of the invisible matter, mainly dark matter (which we'll discuss extensively in Chapter 8), may also be accounted for by this model.

The standard model has so many particles, that in the past some scientists have referred to it as "The Particle Zoo." The standard model predicts the existence of several new fundamental particles. Some of the particles have mass and some don't have mass. Some of these particles have very strange properties, so much so that they are dubbed "strange particles." And indeed, they are strange.

Strange particles are produced copiously, indicating a strong interaction. However, strange particles decay slowly, which indicate a weak interaction. This phenomenon is what originally led to the idea of a weak and strong force. One force is responsible for creating a particle while the other force is responsible for its disintegration.

In the Standard Model, the particles without mass are known as "force mediating particles". The force mediating particles of the

Standard Model are bosons, known as gauge bosons. These gauge bosons are described below:

- Gluons mediating the strong interactions.
- Photons mediating the electromagnetic interactions.
- W and Z bosons mediating the weak interactions.

An exchange of a force mediating particle is equivalent to a force being felt between the two particles. One of the particles predicted by the Standard Model is the "Higgs particle", named after an English theoretical physicist, Peter Higgs. The Higgs particle, sometimes referred to as "The God Particle", is believed to be a force mediating particle that offers an explanation why other particles (except photons and gluons) have mass. Most scientists believe that the reason these particles are so difficult to observe is because of the theorized large energies required to create them. Scientists believe that the experiments at the CERN supercollider may have finally revealed them. They believe they are very close to finally observing them and many scientists believe that they may finally have done so. In Dec, 2011 scientists at CERN believed they collected strong evidence supporting the theory of the Higgs particle:

"We have narrowed down the region where the Higgs particle is most likely to be, and we see some interesting signals, but we need more data before we can reach any firm conclusions," said Physicists, Fabiola Gianotti, who heads the team that works on the collider's enormous Atlas detector. *"It's been a busy time, but a very exciting time."*

On July 4, 2012 scientists at the Large Hadron Collider at the CERN laboratory near Geneva announced the long awaited discovery of the Higgs boson. The announcement came nearly 50 years after it was proposed by physicist Peter Higgs. Higgs along with other scientists came up with the idea of an energy field permeating all of space in 1964. In certain calculations, particles with zero mass kept showing up. Higgs and others realized that once the universe cooled an energy field would have emerged.

The Higgs particles recently measured at CERN had a mass of about 125 billion electron volts (over 130 times the mass of a proton). In order to capture the results, over 500 trillion collisions took place between speeding protons in opposite directions. Out of these trillions of collisions, only a handful produced particles believed to be the elusive Higgs.

The results of the experiment had a statistical significance of about a 1 in 3.5 million chance of an error, convincing most scientists in the validity of the results. There are still some remaining questions to answer. One is whether the data will reveal if the particle's properties match the standard model or if there's a new theory that may be necessary. Another question is where are all the heavy super symmetric partners? There's still much work to be done and perhaps still an open door for new ideas.

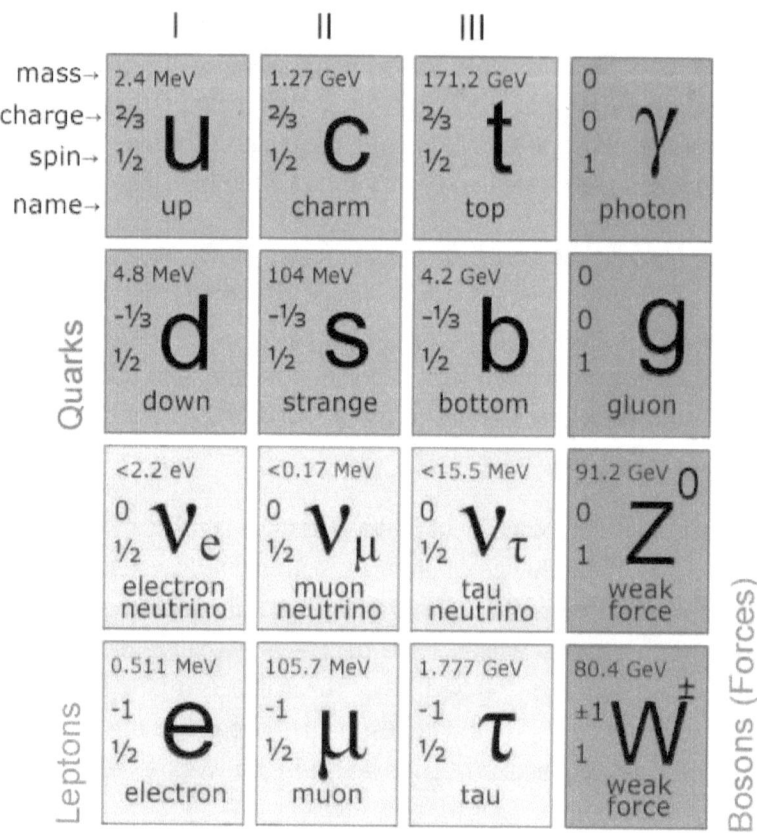

Figure 3: Three Generations of Matter (Wikipedia, 2009)

To help explain what a Higgs particle is a little better, I'd like to present a different way to visualize it:

Imagine that a famous actor steps out of his limousine onto a fancy red carpet at the premier of his smashing new movie, "25 Ways to Prepare and Cook Cats". As he tries to walk to the entrance of the theater, reporters and photographers crowd around him which causes him to slow down. As he slowly makes his way to the door, new people crowd in and take the place of the old people, creating a sort of viscous effect. These pesky people impede his motion and tend to slow him down. In this scenario, these people are equivalent to the Higgs particles. They are mass-less (some would argue also brainless) particles that simply impede motion and thereby give mass to the famous actor, whom without his fans would otherwise be nothing. And yes, his movie was a smashing success.

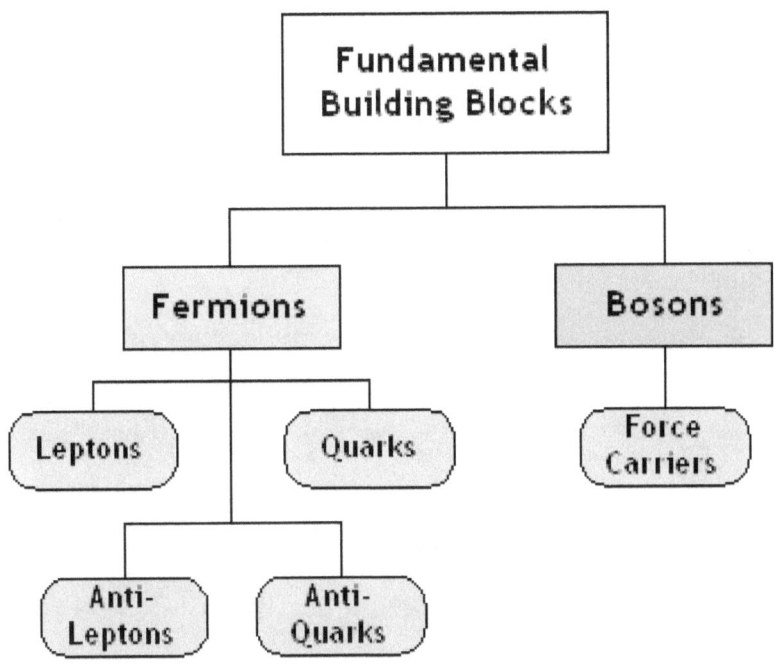

Figure 4: Simplified Diagram of the Fundamental Building Blocks
(http://www.phy.bris.ac.uk, 2009)

Along came Pauli

As stated previously, physicists are hopeful that the Large Hadron Collider (LHC) at CERN (see Figure 5) will reveal or may have already revealed, the Higgs particle and other predicted "super-partners." The Standard Model predicts that for every known particle there is a super-partner in the opposite quantum state. If you don't know much about quantum states, you might surmise that the super-partner for an electron is a proton; after all, they have the opposite charge. This is not the case. Electrons and protons are examples of fermions, that is, they obey "Pauli's Exclusion Principle". Simply put, Pauli's Exclusion Principle states that no two particles can be in the same quantum state at the same time (sometimes we say they can't have the same statistics). The super-partner of an electron must therefore be a boson. In the case of an electron, its super-partner is called a "selectron." Unlike fermions, bosons, do not obey the exclusion principle. Bosons, such as photons, can be in the same quantum state. In fact, they love to be in the same quantum state (at the same time). That's the principle behind how a laser works.

Figure 5: Aerial view of LHC, CERN

In 2011, a major breakthrough with super-partners (sometimes called antiparticles or antimatter) was achieved when Helium's antimatter twin was created at the Relativistic Heavy Ion Collider in Long Island, NY. In this experiment, gold ions were smashed into each other at near the speed of light creating a release of matter and antimatter. With these experiments, physicists have found 18 particles of antihelium^{-4}. This is the antimatter twin of helium, and the heaviest piece of antimatter ever created. This is a remarkable discovery because the more heavy the particle, the less likely it is to form. Some calculations suggest that the probability of two antiprotons and two antineutrons coming together to form antihelium^{-4} is about one in 28 billion. Very low probability odds are common for macroscopic quantum calculations (a bit of an oxymoron to say macroscopic quantum together, so forgive me). For example, one could actually calculate the probability that a tennis ball would tunnel through a brick wall when thrown at it. I did this calculation once many years ago so my numbers are probable way off. But from what I remember, for our minds, the results would make no sense. If you through the ball at the wall each second for a few billion years, the ball may tunnel through once. We're not saying that the ball will break a tunnel through the wall but rather that it will, through quantum probabilities, simply appear on the other side of the wall, just like quantum barrier tunneling for an electron. A helium atom is not even close to the size of a tennis ball, but to be able to achieve these results creating antihelium^{-4} is really amazing.

Forgive me for that very brief description of the Standard Model. While it was a very general overview, it should be sufficient for our purposes.

To summarize this section, the Standard Model predicts that particles with mass and force mediating particles make up all the matter in the universe. Later, we'll discuss another way to possibly understand the unusual Standard Model and the Higgs particle. I'll also try to explain some of the strange quantum observations without using the Standard Model.

Quantum Mechanics Revisited

In Chapter 2, we went over the history and basics of Quantum Mechanics. It was just that, basic. I need to set the stage here, using a bit more detail, so that I can hopefully later make it come crashing down (perhaps not quite so dramatically). It's important to again mention that the observed accuracies in quantum mechanical experiments are very high. These findings are both useful and hard (maybe impossible) to discredit. As I stated before, I don't believe that any experiment has ever been performed to disprove or go against the predictions of quantum mechanics.

Interestingly, Einstein believed that he and a colleague found a hole in quantum theory that couldn't be ignored. Basically, he noted a potential problem in the "Heisenberg Uncertainty Principle", which is one of the cornerstones of Quantum Mechanics. Heisenberg's Uncertainty Principle basically states that the more accurately we know a particle's position the less we know about its speed and momentum (and vice-versa). Using this principle, there is an inherent uncertainty with the information we know about particles. Einstein believed that he and a couple colleagues (Boris Podolsky and Nathan Rosen) finally disproved the Heisenberg Uncertainty Principle with a simple thought problem similar to the following (but not exactly this scenario):

Let's say one particle (particle X) spontaneously decays into two identical daughter particles (particles A and B). To conserve momentum the two daughter particles, with identical masses, will travel in opposite directions, each with the same speed. We could then make a very precise measurement on particle A's velocity. Then we could measure particle B's location and infer its velocity from the measurement we made on particle A. After all it should have the same magnitude, just opposite in direction. We would then know nearly the exact location and speed of both particles. It may not be exactly correct, but the level of precision would exceed those calculated using Heisenberg Uncertainty Principle. We could also know one of the properties of one of the particles without ever making a direct measurement on it.

This was one of many thought problems proposed by Einstein. He would come up with several more that have perplexed scientists for decades. We'll address and answer this prior to the end of the book

Schrödinger's Cat

Another similar and much more famous thought provoking scenario is sometimes referred to as "Schrödinger's Cat". It's so famous that I can't justify not talking about it, especially since it serves as the basis for the title of this book and it involves the possibility of dying cats. Side-note: when I was very young I thought it would be great to be a veterinarian, and then I found out that there was far more to the job then just putting cats to sleep all day.

The scenario goes something like this, written in Schrödinger's words:

"One can even set up quite ridiculous cases. A cat is penned up in a steel chamber, along with the following device (which must be secured against direct interference by the cat): in a Geiger counter, there is a tiny bit of radioactive substance, so small that

perhaps in the course of the hour, one of the atoms decays, but also, with equal probability, perhaps none; if it happens, the counter tube discharges, and through a relay releases a hammer that shatters a small flask of hydrocyanic acid. If one has left this entire system to itself for an hour, one would say that the cat still lives if meanwhile no atom has decayed. The psi-function of the entire system would express this by having in it the living and dead cat (pardon the expression) mixed or smeared out in equal parts."

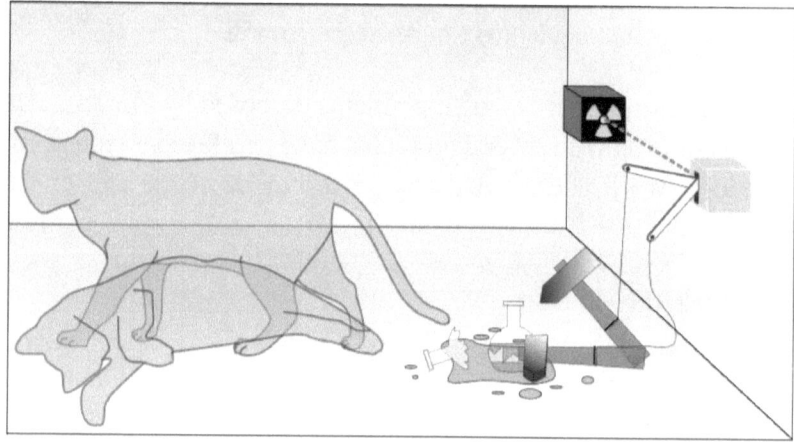

Figure 6: Illustration of Schrödinger's cat (Ard. 2010)

To give a basic summary, this thought experiment states that at any given time, while the box is closed and sealed, the cat is neither dead nor alive. It is rather in some sort of unknown, *blurry* dead-alive state, which coincidentally is probably the same state as most people reading this book.

In these types of thought based scenarios, it's typical that an indeterminacy originally restricted to the atomic domain becomes transformed into macroscopic indeterminacy. This is the incredibly ingenious aspect of this thought problem: the ability to take an atomic probability and have it directly affect a large macroscopic system that we can all understand.

This indeterminacy can then be resolved by direct observation, i.e. opening the box. But in making a direct observation we are said to force an outcome or in mathematical terms we are said to "collapse the wave-function". That observation generally prevents us from accepting as a valid state a "blurred model" for representing reality. In itself, it would not embody anything unclear or contradictory. After all, we see the results. There is a difference between a shaky or out-of-focus photograph and a snapshot of clouds and fog banks. (MacLean. 2009)

I'm not only fascinated about the blurry alive-dead state but I'm also curious as to why a cat was chosen to be in the box. Why not a frog or a dog or a rat or a snail? Where does this morbid fascination about cats stem from? I can only assume that Schrödinger was about as fond of cats as I. His name alone would lead me to believe that he was more of a dog man. But for me, please understand it's not that I don't like cats; in fact, I'm always looking for the perfect recipe.

"It is inexcusable for scientists to torture animals; let them make their experiments on journalists and politicians."
-Henrik Ibsen

I found that Schrödinger's cat thought problem to be so interesting that I wanted to personally verify the experiment. To be certain of the results, I attempted a modified version of this experiment several times, using a camera system inside the box so I could monitor the health of the cats. After many hours and about a dozen cats, I arrived at a surprisingly simple conclusion, and the title for the book: *The cat was dead, sometimes alive.* But I wanted to keep the title positive and cheery, so I chose the dead state. To be certain, there was never a blurry cat lingering between dead and alive states. Sure, a few of them looked rather sick but they were always dead or alive.

Unfortunately, my skeptics will argue that since I was watching the cats with a camera inside the box that this measurement caused the wave-function to collapse. Therefore, my experiment was flawed. On the bright side, however, there are now a few less cats in this unforgiving, dog dominated world.[1]

Einstien Vs. Bohr

Neils Bohr and Einstein argued for years about these and other thought problems designed to discredit quantum mechanics. Bohr was the proponent of quantum mechanics and Einstein died never fully embracing the theory. Many scientists think Einstein wasted the last 20 years of his life attempting to discredit quantum mechanics while searching to find a unified field theory (UFT). I think it's important that we try to thoroughly understand why Einstein never fully accepted quantum mechanics. We may learn something in the process. Plus, I believe we owe that to him, someone who gave so much to science for us.

I find it interesting that Einstein never fully embraced quantum mechanics, since he's known by most scientists as the father of the theory. It was his description of light as a particle, in describing the Photoelectric Effect, which started the quantum revolution. But Einstein did not believe in chance when describing the subatomic world. Bohr, on the contrary, believed that any particle does not occupy a definite space until it is measured. In the case of Schrödinger's cat, in Bohr's world, the cat would be in the dead and alive state at the same time. On a sub-atomic scale, he believed that particles exist solely as a blur of probability.

---

[1] Please forgive my poor attempt at humor. Actually, no *useful* cats were actually harmed during the experiments. In fact, the experiments never even took place.

Einstein also believed that his other theories, general relativity for example, should not break down on the subatomic scale or under a large gravitational field. He believed that the ultimate, or unified theory, should be complete and encompass all of space. He also thought it should contain a touch of elegance. *A Touch of Elegance* was nearly the title of this book but after hours of careful contemplation, I decided to go with the dead cat title. I guess I just liked the sound of it.

Quantum Decoherence

This is a topic that I need to briefly describe if for no other reason than it's the subject of numerous active research projects. A quantum state is believed to be a superposition of other quantum states, for instance, the spin states of an electron. In the "Copenhagen Interpretation", the superposition of states was described by a wave-function, and the wave-function collapse was given the name decoherence.

Today, decoherence researchers study quantum correlations between the states of a quantum system and its environment. But the original sense remains, decoherence refers to the untangling of quantum states to produce a single macroscopic reality. (Schlosshauer, 2005)

In quantum mechanics, quantum decoherence is the mechanism by which quantum systems interact with their environments to exhibit probabilistically additive behavior. Quantum decoherence gives the appearance of a wave function collapse and justifies the framework and intuition of classical physics as an acceptable approximation: decoherence is the mechanism by which the classical limit emerges out of a quantum starting point and it determines the location of the quantum-classical boundary. Decoherence has been a subject of active research since the 1980's.

(Wikipedia, 2009)

Decoherence does not generate an actual wave function collapse. It only provides an explanation for the appearance of wave-function collapse. The quantum nature of the system is simply "leaked" into the environment. A total superposition of the universal wave-function still occurs, but its ultimate fate remains an interpretational issue. Specifically, decoherence does not attempt to explain the problem of measurement. Rather, decoherence provides an explanation for the transition of the system to a mixture of states that seem to correspond to those states we perceive as determinant. But within the framework of the interpretation of quantum mechanics, decoherence cannot explain this crucial step from an apparent mixture to the existence and/or perception of single outcomes. (Wikipedia, 2009) This is an area that I will attempt to address.

Hilbert Space

Analogies are made between classical phase spaces that we can visualize and "Hilbert Spaces". Hilbert space is very commonly used in quantum mechanics and it generalizes the notion of Euclidean space. Hilbert Space extends the methods of vector algebra and calculus from the two-dimensional Euclidean plane and three-dimensional space to spaces with any finite or infinite number of dimensions. A Hilbert space is an abstract vector space possessing the structure of an inner product that allows length and angle to be measured. (Wikipedia, 2009)

A more rigorous derivation in Dirac notation shows how decoherence destroys interference effects and the "quantum nature" of systems.

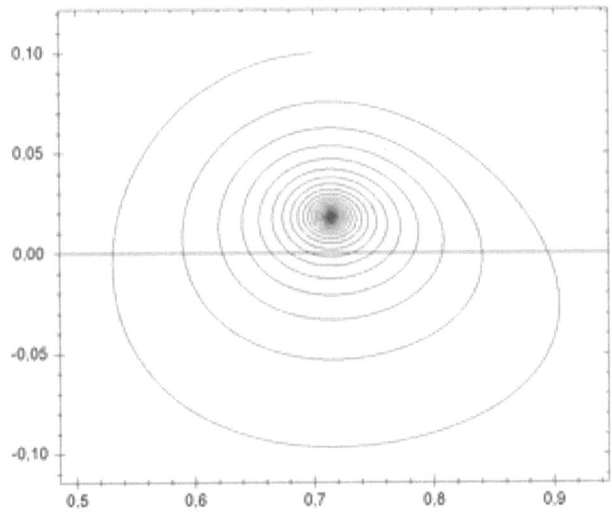

Figure 7: Phase Space Plot

The effective dimensionality of a system's phase space is the number of degrees of freedom present which—in non-relativistic models—is 6 times the number of a system's free particles. When a system couples to an external environment, the dimensionality of, and volume available to, the joint state vector increases enormously. Each environmental degree of freedom contributes an extra dimension.

Any elements that decohere from each other via environmental interactions are said to be "quantumly" entangled with the environment. The converse is not true: not all entangled states are decohered from each other. Any measuring device or apparatus acts as "an environment" since, at some stage along the measuring chain, it has to be large enough to be read by humans. It must possess a very large number of hidden degrees of freedom. In effect, the interactions may be considered to be quantum measurements.

As a result of an interaction, the wave-functions of the system and the measuring device become entangled with each other. Decoherence happens when different portions of the system's wave-function become entangled in different ways with the measuring device. For two elements of the entangled system's state to interfere, both the original system and the measuring system in both elements' devices must significantly overlap, in the scalar product sense. If the measuring device has many degrees of freedom, it is very unlikely for this to happen.

As a consequence, the system behaves as a classical statistical ensemble of the different elements rather than as a single coherent quantum superposition of them. From the perspective of each ensemble member's measuring device, the system appears to have irreversibly collapsed onto a state with a precise value for the measured attributes, relative to that element. This is known as wave-function collapse.

Decoherence represents an extremely fast process for macroscopic objects, since these are interacting with many microscopic objects, with an enormous number of degrees of freedom, in their natural environment. The process explains why we tend not to observe quantum behavior in everyday macroscopic objects. It also explains why we do see classical fields emerge from the properties of the interaction between matter and radiation for large amounts of matter. (Wikipedia, 2009)

A Finite Universe

There are many compelling questions about the history and shape of our universe. After all, what is the shape of the universe and is the universe finite or infinite? And more importantly, why is any of this information important to this book? To tie everything together, these questions must be answered. Theories about whether space is finite or infinite, flat or curved have blazed in the firmament of scientific discourse with varying intensity over time, burning brighter or fading in the face of new data and competing ideas.

I remember once reading about why some ancient civilizations believed the universe was finite. It actually made sense to me at the time. Their logic was quite simple:

First they made the assumption that the stars were distributed fairly uniformly throughout the universe. After all, they had no reason or evidence to suspect otherwise. If the stars were spread out uniformly, and if the universe was infinite then the sum of the brightness of all the stars would make the sky appear very bright, even in the middle of the night (While this sounds logical, I think we would have to see if this sequence of star distribution converges before making this conclusion).

That's obviously not a very scientific argument, but it was quite clever, especially for ancient times. In modern times, our capabilities are much more advanced and sophisticated, thus our understanding is presumably much better. A new study of astronomical data, only recently available, hints at a possible answer to the size and shape of our universe:

This new data suggests that the universe is finite and bears a rough resemblance to a soccer ball or, more accurately, a dodecahedron, a 12-sided volume bounded by pentagons (not quite as pictured in Figure 8).

If this new theory about a soccer ball or dodecahedron shaped universe is proven by further evidence and scrutiny, the model would represent a major discovery about the nature and history of the cosmos.

"What makes it exciting now is it's not a matter of idle speculation," said Jeffrey Weeks, a freelance mathematician in Canton, New York. "There's real data to look at and the possibility of getting a definite answer." (Markey, 2003)

Jeffrey Weeks, was a recipient of a MacArthur Fellowship or the so-called "Genius Award." He spent significant time studying with a team of French cosmologists the cosmic background radiation observed by NASA's Wilkinson Microwave Anisotropy Probe (WMAP). (Markey, 2003)

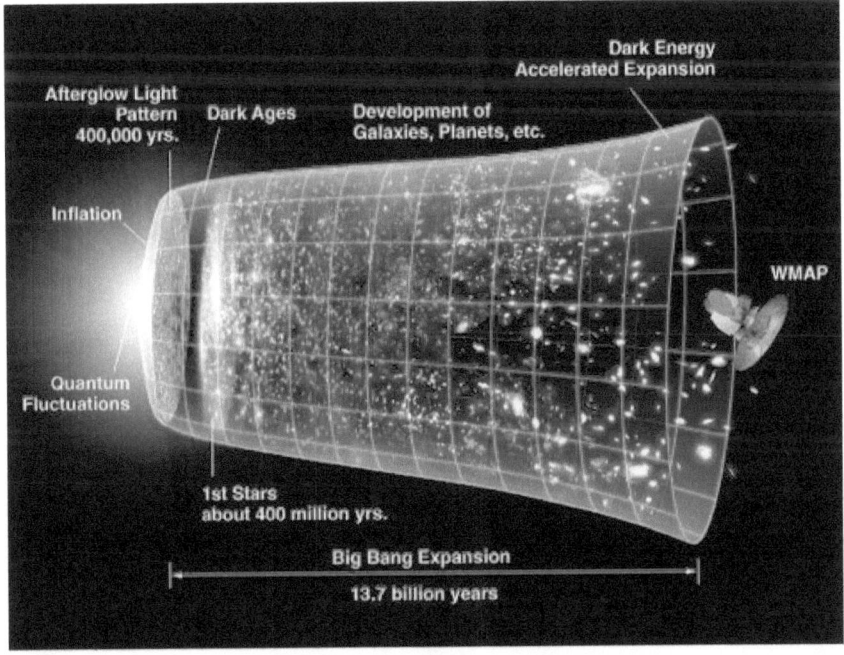

Figure 8: Model of the Universe (Irani, 2008)

With a microwave antenna pointed into deep space and shielded from local interference emanating from the sun, Earth, and moon, a spacecraft has recorded the clearest soundtrack to date of the microwave radiation echo of the Big Bang; the event most scientists believe created the universe. (Irani, 2008)

Where we are Today, *Conclusion*

To conclude the *Where we are Today* chapter, I'd like to again point out that there's much more to understand. In the special and general relativity chapter to follow, we'll add to our understanding and start to travel down new roads.

For the remainder of this book, I want to approach each unexplained physical phenomena from an Einsteinium and sometimes a Newtonian point of view. Approaching modern problems this way, will certainly diverge from quantum and modern day theory. The results will not always be correct. But the final goal is to understand and be able to explain what's happening on the subatomic and macroscopic scales and perhaps develop a new model for the universe.

Special & General Relativity

"Put your hand on a hot stove for a minute, and it seems like an hour. Sit with a pretty girl for an hour, and it seems like a minute. That's relativity." Albert Einstein

What's so *Special* about Relativity?

I wish the opening quote by Einstein was sufficient to explain relativity, but I'm afraid it's not. However, a detailed explanation on relativity is really not much more complicated than that. So that's how we'll approach it in this chapter. Basically, in the opening quote by Einstein, he suggests that time does seem to pass differently at different times. I think that most people have observed this phenomenon in their own lives. That's relativity in a nutshell; certainly it's not worth being intimidated over, especially at our level in this chapter.

Having a small background about relativity will be necessary to understand the subatomic world. That's the main reason that we'll at least attempt to understand it. Regarding the subatomic world, relativity is so important that I wanted to devote an entire chapter to the subject. This should be considered kind of the transitional chapter in this book. It's the page that separates the New from the Old Testament. In this chapter we'll begin to move away from what we think we know, and we'll start to explore new ideas and new possibilities.

Many scientists, far smarter than me, have spent decades trying to link the subatomic world to gravity. This is far more difficult than it may at first appear. If we recall from the previous chapter, the strong force is approximately 10^{38} times stronger than the gravitational force. How in the world can we unite these forces? That's the essence of a unified field theory (UFT) or sometimes called "a theory of everything." *Despite all the failed attempts in the past, I'm convinced that the search for a unified theory is not a total waste of time.*

Nowadays, unified field theories are a dime a dozen. One estimate stated that there are over 10^{100} possible solutions to the "field equations". But we need to keep in mind that these solutions are just mathematical interpretations to a problem that don't offer us any real understanding. In this chapter and for the remainder of the book, I want to try to rationalize and understand a solution, not simply search for an answer hidden in mathematical puzzle. This will be a challenging, perhaps impossible, feat.

Based on our current understanding of science and physics, forces between objects (e.g. gravitation) are not believed to be transmitted directly between the two objects. Instead the forces are believed to go through intermediary entities called fields. All four of the known fundamental forces are believed to be mediated by fields, which in the standard model of particle physics result from exchange of bosons, as previously discussed in Chapter 3. Specifically the four interactions or forces to be unified are the strong force, weak force, electromagnetic force and gravitational force.

The goal of a modern unified field theory is to bring these four force-mediating fields together into a single framework. Quantum theory seems to limit any deterministic theory's descriptive power.

In simple terms, this means that it is believed that no theory can predict events more accurately than allowed by the Planck constant. (Wikipedia, 2009)

That's just one of the conflicts with unifying the four fundamental forces. To understand some of the other challenges, we first need to understand gravity and the disparities between the weak gravitational force and the strong electromagnetic force. To do this, I'd like you to consider the following simple story:

A cat jumped out of an airplane without a parachute (probably not the wisest decision this feline ever made). The gravitational force caused the cat to travel about 15,000 feet where it smashed into a road and was stopped by the electromagnetic repulsive force from the atoms in the concrete within a few inches (May it rest in peace).

Like many of the examples in this book, this too may seem quite trivial and useless, but I can assure you that it's not. In fact, Albert Einstein had a similar idea which he called, "The Happiest Thought of my Life." The simple question was asked:

What happens when a roofer falls off the roof? What happens? He goes splat! —and that's where most of us, including most scientists, would leave the image. However, Einstein did not. He focused on the time right before the roofer hit the ground. Almost immediately, Einstein realized that as the roofer fell, the man would not feel his own weight. To Einstein, that meant that the experience of being in free fall in a gravitational field and that of floating in a gravity-free region of space were effectively the same, or equivalent. That formed the basis of what he called the "Equivalence Principle", which serves as the foundation of the General Theory of Relativity. We will discuss the Equivalence

Principle in much greater detail later on in this chapter. For now we will keep this idea brief. (Levenson, 2004)

In 2010, I watched an interesting movie, or at least I dreamt that I did, called *Inception*. In this movie, they show a wonderful way that we can use to visualize the Equivalence Principle. I'm pretty sure this was not the director's intention but if you've seen this movie it may help to understand this principle. If you haven't seen the movie this will make little to no sense, so you should just skip to the next paragraph and watch the movie at your earliest convenience. In the movie *Inception*, there's a scene that has people in a dream within a dream. In the deeper of the two dreams, the characters are able to float or fly as if they are weightless. In the less deep dream, they are in free-fall, falling off a bridge in their car. The free-fall from one dream is equivalent to being weightless in the deeper dream. It was quite remarkable to see this on the big screen in a huge Hollywood production. From what I saw, it was also quite an engineering feat to make this scene come to life, and a select few of us appreciated the scientific accuracy that was crafted into it. It's a great improvement from the days of the physics defying road runner and coyote scenes that I grew up watching.

I think that it's amazing how much can be learned from performing or observing simple manual labor. Who would have ever thought that a roofer would contribute so much to science? Basic thought questions like this were very intriguing and very important to Einstein. In fact, he used another thought question which helped him to begin to develop his famous Special Theory of Relativity. Einstein asked the question, "What happens if you try to catch up to a beam of light?" His actual question involved looking into a mirror while speeding up approaching the speed of light. What happens to the light coming off your face? Will you appear to be invisible while looking into the mirror?

The answer surprised almost everyone. You see, Einstein realized that the laws of electromagnetism could not depend on relative motion and therefore, the speed of light must also be the same no matter how the observer or source moves. Since light is a propagating electromagnetic wave, there must have been a flaw in the theory of the time. You see, if you were traveling at the speed of light and observed a photon (recall that a photon is a single particle of light) there would be no wave motion. Since there would be no wave motion, the electric and magnetic fields could not support themselves, therefore there would be no photon. As a result, *Einstein concluded that the speed of light is the universal speed limit for objects with mass.*

He was then able to fix the speed of light and work out his famous results using Maxwell's equations. By doing this, he was able to show that mass and energy are equivalent and he derived perhaps his most famous equation: $E=mc^2$. Using this idea, he also showed that space were time are not fixed and were far different than previously believed. (Smolin, 2009)

These ideas laid the groundwork for what became known to the world as "Special Relativity." Special Relativity shocked the scientific community, mainly because of its very interesting consequences. The first unusual consequence of Special Relativity, simply put, is that time passes differently in different reference frames. In Special Relativity, time and space were no longer fixed. A classic example, one that's most commonly used, to help describe this, is known as the *"Twin Paradox"* (shown in Figure 9).

Figure 9 shows that one twin leaves earth traveling very fast and returns home to be reunited with his twin who remained on earth. When he gets home, he notices some very interesting differences in his brother.

Twin Paradox

one set of twins leaves the Earth in a rocketship bound for the stars

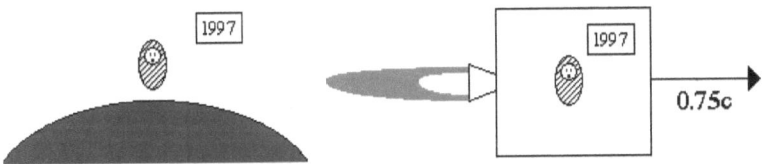

60 years later the rocket returns to with the astronaut only 40 years old due to time dilation

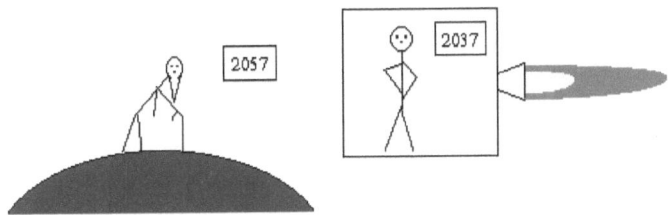

Figure 9: Diagram of the Twin Paradox (http://www.oglethorpe.edu, 2010)

They say that a picture is worth a thousand words, well if that's true the one above is worth a thousand and one. This classic example is actually what first got me hooked on science. Basically this picture shows that one twin travels at very close to the speed of light (0.75c) away from the earth and returns 60 earth years later. When he returns to earth, he notices that his twin brother has aged 60 years while he, the space traveling twin, has only aged 40 years. Imagine that!

Now if you're like me, you might be thinking, "Could this be a way to extend our lives or possibly live forever?" Wouldn't that be great? Don't waste too much time with this thought because, I'm afraid it's not possible. You see, special relativity shows us that for the space traveling twin, he thought he left earth for only 40 years, not 60 years. To him, time still passed and he still aged, just slower than his brother on earth. And in reality, the earth twin probably had much more fun than the brother who was cooped in a rocket for decades.

You may be wondering why they call this scenario the *Twin Paradox*. After all, what's the paradox here? The paradox of this problem is very subtle and it is not really a paradox at all. It's really just a matter of perspective. If you're on earth, you see the rocket ship leave and then come back to earth. But for the space traveler, assuming he is watching earth as he leaves, the earth twin and the earth appear to be moving near the speed of light away from him. When he leaves in his very fast rocket ship, it looks like the earth is moving away near the speed of light. After all, their movement and velocity are also relative. So why then does the earth twin age more rapidly? After all, most scientists believe there is no absolute fixed reference frame by which we can measure different velocities. This idea may or may not be true; we'll return to it in greater detail later.

This paradox is usually resolved by noting that the earth traveler must reverse directions to return to earth while the earth twin does not have to do anything. In 1911, a famous scientist, Paul Langevin, stated, *"Only the traveler has undergone an acceleration that changed the direction of his velocity."* According to Langevin, acceleration here is "absolute", in the sense that it is the cause of the asymmetry (and not of the aging itself). (Langevin, 1911)

As shown by another famous scientist, Max von Laue, in 1913, the process of acceleration is not as important as Langevin suggested. This is because the asymmetric aging is completely accounted for by the fact that the astronaut twin travels in two separate frames, while the earth twin remains in one frame. Using Minkowski's space-time formalism (shown in Figure 10), Laue went on to demonstrate that the world lines of the inertially moving bodies maximize the proper time elapsed between two events. (Laue, 1913)

The significance of the *Twin Paradox* hinges on this one crucial detail of asymmetry between the twins. It should be stressed that neither Einstein nor Langevin considered such results to be literally paradoxical: Einstein only called it "peculiar" while Langevin presented it as evidence for absolute motion.

Resolution of Twin Paradox in Special Relativity

Generally, the standard textbook description treats the twin paradox as a straightforward application of special relativity. Here the Earth and the ship are not in a symmetrical relationship: the ship has to "turnaround" in which it undergoes non-inertial motion, while the Earth has no such turnaround. Since there is no symmetry, it is not paradoxical if one twin is younger than the other. Nevertheless, I think it's still useful to show that special relativity is self-consistent, and to show how the calculation is done from the standpoint of the traveling twin (the calculation here will only be qualitatively described).

Special relativity does not claim that all observers are equivalent, only that all observers at rest in inertial reference frames are equivalent. But in this scenario, the space ship jumps frames (accelerates) when it performs a U-turn. It also accelerates when it leaves earth and decelerates when it slows down when it returns home. But if we were to consider equal and opposite forces, there will be a slight acceleration on the earth when the rocket takes off. So it's easier and still fairly accurate to ignore when the ship takes off and returns and just focus on the turn-around point of the space traveling twin. It should be obvious that the twin who stays at home remains in the same inertial frame for the whole duration of his brother's flight. No accelerating or decelerating forces apply to the homebound twin, because there's no turn around.

To be more precise, there are actually three relevant inertial frames to consider:

1. The inertial frame in which the stay-at-home twin remains at rest.

2. The inertial frame in which the traveling twin is at rest on his outward trip.

3. The inertial frame in which he is at rest on his way home.

It's during the acceleration at the U-turn that the traveling twin switches reference frames. That's when he must adjust his calculated or perceived age of the twin at rest.

In special relativity there's no concept of absolute present. A present is defined as a set of events that are simultaneous from the point of view of a given observer. The notion of simultaneity depends on the frame of reference, so switching between frames requires an adjustment in the definition of the present. If one imagines a present as a (three-dimensional) simultaneity plane in Minkowski space, then switching frames results in changing the inclination of the plane.

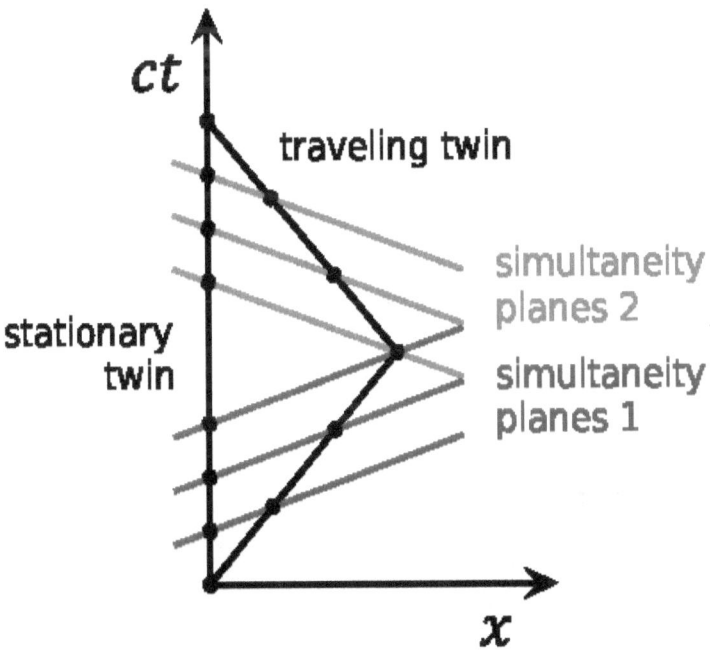

Figure 10: Minkowski Space-Time Diagram (Wikipedia, 2010)

In the space-time diagram above, drawn for the reference frame of the stay-at-home twin, the stay-at-home twin's world line coincides with the vertical axis (his position is constant in space, moving only in time). On the first leg of the trip, the second twin moves to the right (black sloped line); and on the second leg, back to the left. The 3 parallel, positively slopped lines show the planes of simultaneity for the traveling twin during the first leg of the journey; the 3 parallel, negatively slopped lines, during the second leg. Just before the turn around, the traveling twin calculates the age of the resting twin by measuring the interval along the vertical axis from the origin to the upper positively slopped line. Just after turnover, if he recalculates, he'll measure the interval from the origin to the lower negatively slopped line.

In a sense, during the U-turn the plane of simultaneity jumps from the positively to negatively slopped lines and very quickly sweeps over a large segment of the world line of the resting twin. The traveling twin reckons that there has been a jump discontinuity in the age of the resting twin. In this picture, the traveling twin thinks that only 6 line segments of time has passed (only 6 dots are travelled through) while the stationary twin thinks that 7 time segments have passed. That's another way to reason the apparent difference in ages of the twins. (Wikipedia, 2009)

Why Time Passes Differently

One of the features of Special Relativity is that it shows us a new way to compute and to think of time. But the theory does little to help us understand the million dollar question: *why?* For me, the underlying reason is far more important than the result.

I want to start this section about the passing of time by considering a slightly different version of the twin paradox. I'll call it, *"The Alien Comet"* (see Figure 11). Picture this scenario:

An alien, from a distant galaxy, is stranded on a run-away comet headed straight towards earth. Astronomers, in an observatory with a clock mounted on the top of their station, in the United States, estimate the comet is one light year away (about 9.5 trillion kilometers). They further conclude that the comet is traveling very close to the speed of light (.99c). The scientists therefore conclude that the comet will strike the earth in approximately 1 earth year. The alien on the comet is equipped with the finest viewing and time measuring equipment that alien engineering can produce. The alien and scientists all decide to monitor each other's clocks. How many earth years does the alien expect before impact? What do the alien and scientists observe when monitoring their counterpart's clocks?

Figure 11: Comet heading towards earth at .99c with a time-keeping alien observer

This is very similar to the *Twin Paradox* scenario except there's no change in direction (thus no acceleration; only a deceleration before it crashes). Also, both observers are able to watch each other for the entire journey.

There are many problems with this scenario that I don't want to focus on, so I'll still briefly point them out. First, I don't think it's realistic to say that an alien can travel at .99c. After all, we don't have any concrete proof that aliens exist (plus they probably couldn't survive that speed). Also, when traveling very fast we need to consider certain effects. One such effect is the perceived direction of light rays due to our motion. This is known as "aberration". At very high speeds, close to the speed of light, this effect cannot be ignored. Basically, if we move in some direction, then as compared to what we see at rest, we see that the light rays arrive more from the front. This has several effects, the most dramatic of which is that we will see in front of us things that we have already physically passed. This is very difficult to visualize and comprehend but it's a real effect.

We'll also have to contend with a large shift in the frequency of the incoming light. At these speeds, objects may appear as simply a blur of ultraviolet radiation. The object may be so blurry that we wouldn't be able to distinguish anything. But I'm going to ignore these effects and attempt to answer the original questions.

First, there are a few things we know. Using the formulas from Einstein's Special Theory of Relativity, the relativistic factor for a speed of .99c is about 7. You can trust me on the math. It may not be exact but it is close enough. Simply put, a relativistic factor of 7 means a few things, specifically:

1) The comet's mass, and the alien's mass, would appear to increase by a factor of 7.

2) The comet and the alien would contract in the direction of travel by 7, meaning a 700 foot comet would appear to shrink to about 100 feet (to the stationary observer on earth).

3) Additionally, due to time dilation, one year to the astronomers on earth would seem to be about 52 days for the alien on the comet.

We also know that information travels to each observer at a relative speed of c. I'd like to now make an assumption:

Assumption: No information is lost between observers.

Basically, this assumptions means that all movements of both clocks are observed by both parties. I've struggled with that assumption for some time now, and I'll address this in a later chapter. It may seem basic, but this assumption may have huge implications. So far, I've only made a couple assumptions in this book that I'll recap here:

1. *There must be a point where we cannot divide particles any further.*

2. *No information is lost between observers in relative motion with each other.*

These two assumptions do not go well together at all. In the first assumption, *I assume that there's a smallest unit of matter or mass.* The second assumption states that even though time may pass differently between observers the information exchanged between them remains intact (i.e. nothing is lost).

If no information is lost, when both parties are looking at their counterpart's clock, what will they each observe? Surprisingly, both observers will think their counterpart's clock is running slow. So how can this be possible?

Let's assume both parties observe 12:00 on their counterpart's clock at the same time. I don't like using phrases like *"at the same time"* because entirely new issues have to be addressed when I do this. But bear with me for a moment. At a later time, the alien observes 1:00 on the earth clock but he reads 1:08 on his clock. At the same instance, the astronomers observe 1:00 on the alien's clock and 1:08 on their own clock. Both observers think their counterpart's clock is running slow!

Many people would think the obvious answer to the original question about what each party observes would be simple. At first, most would think the alien would observe the earth clock moving abnormally fast and the astronomer would observe the comet's clock moving in slow motion. I think many people would assume this because it's the alien who is barreling to the earth. But the problem with this solution is that both parties remain in their original reference frame (no acceleration), so this cannot be the case. Plus, even though the earth appears still in the example, *there is no concept of absolute rest.*

Now when the comet crashes into the earth, both clocks will read the same time, as they must. You see, as the comet gets closer to earth, each observer will notice the difference in time, the delta, between the clocks to be getting smaller. So in effect they are becoming synchronized. This is due to the fact that the light has a shorter distance to travel as the comet approaches closer and closer to earth.

Not So General Relativity

General Relativity is not as difficult as most people make it out to seem. It also has an interesting history behind it. I was surprised when I found out that shortly after Einstein published his *Special Theory of Relativity*, he began to dislike it very much. In fact, it was so soon after, that most of the scientific world was just beginning to embrace and understand it (as much as they were capable of understanding). At the time, Einstein's dissatisfaction surprised to the entire scientific community. There were a couple reasons why Einstein disliked his new theory so much, specifically:

- First, special relativity did not hold for all observers. Specifically it did not hold for those in an accelerating reference frame.

- Secondly, he thought it was necessary to be able to incorporate gravity into his theory, making use of the Equivalence Principle.

This is really where the Equivalence Principle comes into the picture. Einstein discussed the Equivalence Principle in a recall of events of 1907 in a talk in Japan on 14 December 1922 (Morikawa, 2005) and stated:

"A little reflection will show that the law of the equality of the inertial and gravitational mass is equivalent to the assertion that the acceleration imparted to a body by a gravitational field is independent of the nature of the body. For Newton's equation of motion in a gravitational field, written out in full, it is:

(Inertial mass) (Acceleration) = (Intensity of the gravitational field) (Gravitational mass)

It is only when there is numerical equality between the inertial and gravitational mass that the acceleration is independent of the nature of the body."

When developing his theory, Einstein stated:

we [...] assume the complete physical equivalence of a gravitational field and a corresponding acceleration of the reference system. -Einstein, 1907

More simply put: *being at rest on the surface of Earth is equivalent to being inside a spaceship (far from any source of gravity) that is being accelerated by its engines.*

From this principle, Einstein deduced that free-fall is actually inertial motion. Objects in free-fall really do not accelerate, but rather the closer they get to an object such as the Earth, the more the time scale becomes stretched due to space-time distortion around the planetary object (this is gravity). An object in free-fall is in actuality inertial, but as it approaches the planetary object the time scale stretches at an accelerated rate, giving the appearance that it is accelerating towards the planetary object when, in fact, the falling body really isn't accelerating at all. This is why an accelerometer in free-fall doesn't register any acceleration; there isn't any.

By contrast, in Newtonian mechanics, gravity is assumed to be a force. This force draws objects having mass towards the center of any massive body. At the Earth's surface, the force of gravity is counteracted by the mechanical (physical) resistance of the Earth's surface. So in Newtonian physics, a person at rest on the surface of a (non-rotating) massive object is in an inertial frame of reference. These considerations suggest the following corollary to the equivalence principle, which Einstein formulated precisely in 1911:

Whenever an observer detects the local presence of a force that acts on all objects in direct proportion to the inertial mass of each object, that observer is in an accelerated frame of reference.

Einstein also referred to two reference frames, K and K'. K is a uniform gravitational field, whereas K' has no gravitational field but is uniformly accelerated such that objects in the two frames experience identical forces:

We arrive at a very satisfactory interpretation of this law of experience, if we assume that the systems K and K' are physically exactly equivalent, that is, if we assume that we may just as well regard the system K as being in a space free from gravitational fields, if we then regard K as uniformly accelerated. This assumption of exact physical equivalence makes it impossible for us to speak of the absolute acceleration of the system of reference, just as the usual theory of relativity forbids us to talk of the absolute velocity of a system; and it makes the equal falling of all bodies in a gravitational field seem a matter of course. -Einstein, 1911

This observation was the start of a process that culminated in General Relativity. Einstein suggested that it should be elevated to the status of a general principle when constructing his theory of relativity:

As long as we restrict ourselves to purely mechanical processes in the realm where Newton's mechanics holds sway, we are certain of the equivalence of the systems K and K'. But this view of ours will not have any deeper significance unless the systems K and K' are equivalent with respect to all physical processes, that is, unless the laws of nature with respect to K are in entire agreement with those with respect to K'. By assuming this to be so, we arrive at a principle which, if it is really true, has great heuristic importance. For by theoretical consideration of processes which take place relatively to a system of reference with uniform acceleration, we obtain information as to the career of processes in a homogeneous gravitational field. -Einstein, 1911

Einstein combined (postulated) the Equivalence Principle with Special Relativity to predict that clocks run at different rates in a gravitational potential, and light rays bend in a gravitational field, even before he developed the concept of curved space-time.

So the original Equivalence Principle, as described by Einstein, concluded that free-fall and inertial motion were physically equivalent. This form of the Equivalence Principle can be stated as follows:

An observer in a windowless room cannot distinguish between being on the surface of the Earth, and being in a spaceship in deep space accelerating at 1g. (This is not strictly true, because massive bodies give rise to tidal effects, caused by variations in the strength and direction of the gravitational field, which are absent from an accelerating spaceship in deep space)

Although the Equivalence Principle guided the development of General Relativity, it is not a founding principle of relativity but rather a simple consequence of the geometrical nature of the theory. In General Relativity, objects in free-fall follow geodesics of space-time, and what we perceive as the force of gravity is instead a result of our being unable to follow those geodesics of space-time, because the mechanical resistance of matter prevents us from doing so.

Before an experiment could even be performed, Einstein, published his *General Theory of Relativity.* Later, in 1919, a rare opportunity presented itself and his General Theory of Relativity was confirmed, making Einstein a scientific superstar.

Like Special Relativity, General Relativity also presented a whole new way to view the universe. With

"Gravitation is not responsible for people falling in love."
- Albert Einstein

General Relativity, space and time are not fixed but rather they continuously evolve dynamically. When most people learn of General Relativity, they learn the basics, mainly curved space and the bending of light.

Under a large gravitational field (say caused by a star), space is distorted causing light and other objects to bend around the star, as shown in Figure 12.

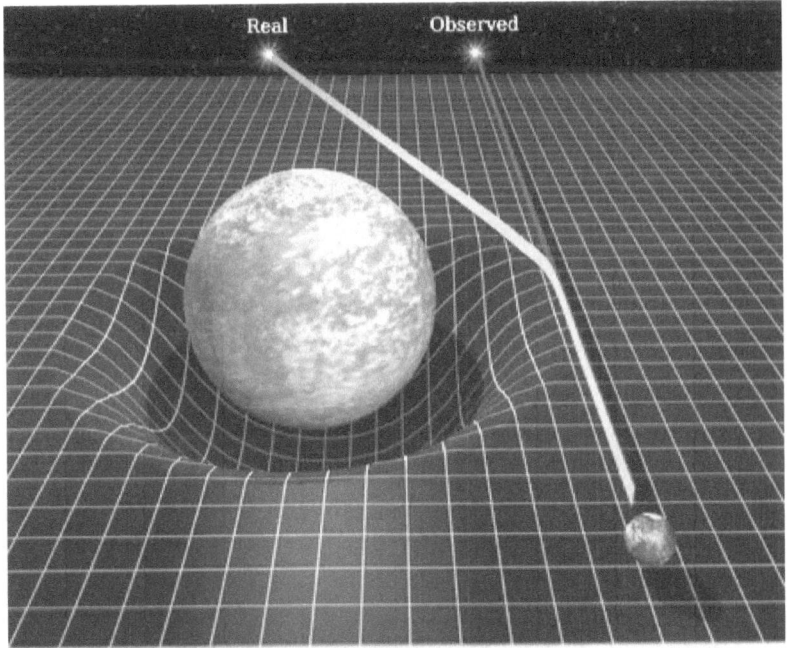

Figure 12: bending light around the sun due to curved space
(http://thetechies.files.wordpress.com, 2010)

Figure 12 shows the basics of the now famous experiment conducted in 1919 where light from a distant star was able to be observed during a solar eclipse. Under normal circumstances, the light should not have been able to be seen. The bending of the light around the sun due to curved space essentially proved General Relativity.

General Relativity is, however, much more complicated than curved space, bending light, the big bang theory, and even black holes. It requires a whole new way of thinking in order to understand the universe. To help to understand and appreciate General Relativity, I usually think of it as a theory about the general behavior of space and time.

Prior to the 20th century, scientists formed a basic background to attempt to describe the movement of matter. The role of a physical theory was to describe how different kinds of matter interacted with one another and to be able to predict their motions. With the development of General Relativity, the role of space and time in the theories of physics were changed dramatically. Instead of a fixed background, space and time became viewed as dynamic variables. Scientists began to believe that space and time were capable of changing by the matter within them and in turn they changed the way that matter behaves.

In General Relativity, space-time becomes curved because of the effects of matter. Basically, in curved space-time the laws of Euclidean geometry no longer hold. The specific laws that are not always true in General Relativity are:

The angles of a triangle do not always add up to 180°. Additionally, the ratio of the circumference of a circle to its diameter is not always pi, etc...

As a result of this curvature, the behavior of matter is affected. In Newtonian physics, a particle in motion with nothing pushing or pulling it (no forces acting on it) will move in a straight line. In curved space-time, the lines are now twisted and bent, and particles with no forces acting on them are seen to move along curved paths.

To help explain this with our planet, I have a simplistic view of General Relativity. This view helps me to rationalize elliptical orbits, gravity and *Kepler's Second Law of Planetary Motion.*

Kepler's Second Law of Planetary Motion states that a line joining a planet and the Sun sweeps out equal areas during equal intervals of time (Bryant, Pavlvk, 2009). The connection that I'm about to explain between Kepler's Laws and General Relativity may be completely wrong, but nonetheless I'll attempt to describe it.

Kepler's Laws & General Relativity – *New Idea*

Einstein described that the mass of a planet creates a dent in space-time, similar to placing a heavy ball across a sheet of stretched out rubber, similar to Figure 13. Let's assume this analogy to be true. If you could look straight down on that planet (depicted as earth in Figure 13), the dent in space-time would look like a circle. Now imagine drawing a line around the space-time curve, equal in height at all points to the center of the planet. This too would be a circle. Figure 13 shows this circular path, denoted by the bright color line.

Figure 13: Space-time curve depicting a circular path in bright color (Bryant, Pavlvk, 2009)

Now imagine we shift the starting point of the line lower than the center of the planet and draw around the space-time curve so the highest point is above the center of the planet. If done correctly, this could form an ellipse, viewing from straight down. Figure 14 shows this elliptical path, again denoted by the bright color line.

Figure 14: Space-time curve depicting an elliptical path in the brighter color

Now imagine that a satellite in orbit follows this path:

When the satellite is at its lowest point, it is deepest in the gravity well and will therefore travel the fastest. I relate this to a marble rolling around in a circular bowl. At the lowest point in the well, the satellite is also closest to the planet. At the highest point in the orbit, the speed will be the slowest. At this point it is farthest from the planet. This description nicely corresponds and agrees with Kepler's Law.

As stated above, this can be viewed as being similar to rolling a ball around a bowl. At the bottom of the bowl, the speed (kinetic energy) is the fastest and at the top of the bowl the speed is the slowest (potential energy is the greatest). This causes objects to speed up when they are close to a large body and to slow down when they are farther away from a large body. The object is closest to the large body at lowest point in the potential well. Here the potential energy is lowest while the orbiting speed is the highest.

Note: it is far more complicated than this simplified example because the orbiting body also creates a dent in space and will also move around in this well as it orbits around the large body in the large body's well.

Kepler's 2ⁿᵈ Law of Planetary Motion states that the Sun sweeps out equal areas during equal intervals of time. The explanation of orbits shown in Figure 15 below can be conveniently explained using the large bowl and ball analogy presented earlier. Figure 15 is a highly exaggerated elliptical orbit. In reality, the orbit of the earth around the sun differs little from a circle.

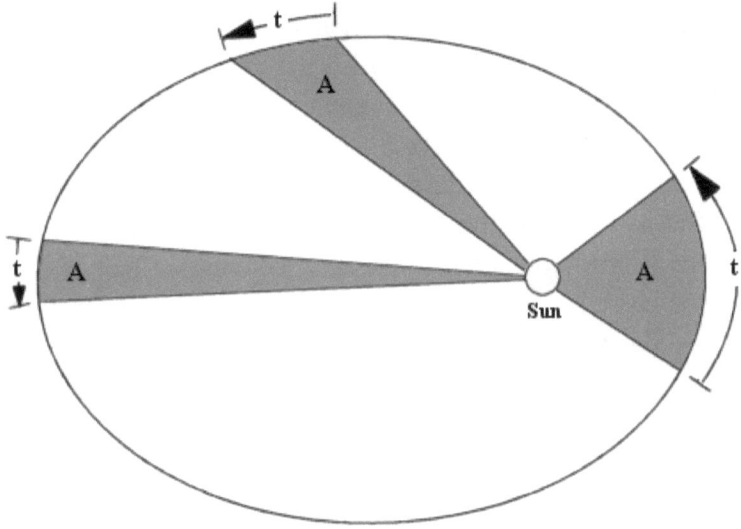

Figure 15: Diagram of Kepler's 2nd Law (Wikipedia, 2010)

The bowl and ball analogy with energy wells can also be used to explain elliptical orbits. This new idea or interpretation could, of

SPECIAL & GENERAL RELATIVITY

course, be completely wrong. But regardless, this explanation offers an easy way to visualize and remember *Kepler's Laws of Planetary Motion.*

Resolution of Twin Paradox in General Relativity

Regretfully, we're back to the twin paradox scenario for a moment, but not for long I can assure you. Basically I want to try to understand how (if at all) the twin paradox can be resolved in General Relativity. The issue or point on contention in the General Relativity solution is how the traveling twin perceives the situation during the acceleration for the turn-around. This issue is well described, much better than I can do, in Einstein's twin paradox solution of 1918. (Einstein, 1918)

In this solution, it was noted that from the viewpoint of the traveler, the calculation for each separate leg equals that of Special Relativity, in which the Earth clock's age is less than the traveler's clock. For example, if the Earth clocks ages one day less on each leg, the amount that the Earth clocks will lag behind due to speed alone amounts to two days. Now, the accelerated frame is regarded as truly stationary, and the physical description of what happens at turn-around has to produce a contrary effect of double that amount: in other words, four days advancing of the Earth's clock. For this reason, the space traveler's clock will finish up with a two day delay on the Earth clocks, just as Special Relativity stipulates.

The mechanism for the advancing of the stay-at-home twin's clock is gravitational time dilation. When an observer finds that inertially moving objects are being accelerated with respect to themselves, those objects are in a gravitational field insofar as relativity is concerned. For the traveling twin at turn-around, this gravitational field fills the universe. (It should be noted and emphasized that according to Einstein's explanation, this gravitational field is just as "real" as any other field, but in modern interpretation it is only perceptual because it is caused by the traveling twin's acceleration).

In a gravitational field, clocks tick at a rate of $t' = t(1 + \Phi / c2)$

Φ is the difference in gravitational potential. In this case, $\Phi = gh$

g is the acceleration of the traveling observer during turnaround

h is the distance to the stay-at-home twin

h is a positive value in this case since the rocket is firing towards the stay-at-home twin thereby placing that twin at a higher gravitational potential. Due to the large distance between the twins, the stay-at-home twin's clocks will appear to be sped up enough to account for the difference in proper times experienced by the twins. It is no accident that this speeding up in clock time is enough to account for the simultaneity shift previously described above.

Although this is called a "General Relativity" solution, in fact it is done using the findings related to Special Relativity for the accelerated observers that Einstein described as early as 1907 (namely the equivalence principle and gravitational time dilation). For this reason it could be called the "accelerated observer viewpoint" instead. It can be shown that the General Relativity solution for a static homogeneous gravitational field and the Special Relativity solution for finite acceleration produce identical results. (Jones, 2006)

String Theory

Due to the popularity in mainstream media, I felt compelled to discuss String Theory or Superstring Theory at some point in this book, even though I'm certainly no expert on the subject. Most of this section is from the work of some of the leading researchers in the field of string theory with much of their work available for viewing at superstringtheory.com.

What better place to talk about string theory than in the Special and General Relativity chapter. Seems a little out of place, right? It may not seem right to describe it here, but I have my motivations for approaching it in this fashion.

There have been several attempts at unifying the fundamental theories into a *Grand Unification Theory (GUT)*. Currently, many scientists think that string theory offers the most promise. So what is string theory? A basic way to visualize it is to think of a guitar string that has been tuned by stretching the string under tension across the guitar. Depending on how the string is plucked and how much tension is in the string, different musical notes will be created by the string. These musical notes could be said to be excitation modes of that guitar string under tension.

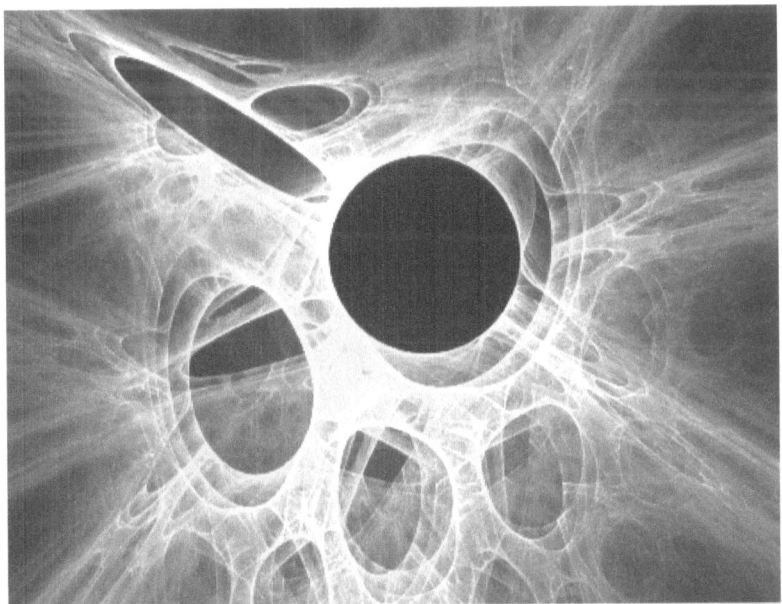

Figure 16: Artist's View of Super Strings (http://larvalsubjects.wordpress.com, 2010)

In a similar manner, in string theory the elementary particles we observe in particle accelerators could be thought of as the musical notes or excitation modes of elementary strings. This is vastly different than how most people currently visualize the elementary particles.

In string theory, as in guitar playing, the string must be stretched under tension in order to become excited. However, the strings in string theory are floating in space-time. They aren't tied down like on a guitar. Nonetheless, they are believed to have tension.

If string theory is to be a theory of quantum gravity, then the average size of a string should be somewhere near the length scale of quantum gravity, called the *Planck length*, which is about 10^{-33} centimeters, or about a millionth of a billionth of a billionth of a billionth of a centimeter. Unfortunately, this means that strings are way too small to see by current or perhaps even by future particle physics technology. Therefore, string theorists must devise more clever methods to test the theory besides just looking for little strings in particle experiments.

String theories are classified according to whether or not the strings are required to be closed loops, and whether or not the particle spectrum includes fermions. In order to include fermions in string theory, there must be a special kind of symmetry called super-symmetry, which means for every boson (as previously described, these are particles that transmits a force) there is a corresponding fermion (particles that makes up matter). So super-symmetry relates the particles that transmit forces to the particles that make up matter.

Super-symmetric partners to currently known particles have not been observed in particle experiments, but theorists believe this is because super-symmetric particles are too massive to be detected at current accelerators. Particle accelerators could be on the verge of finding evidence for high energy super-symmetry in the next decade. Evidence for super-symmetry at high energy would be compelling evidence that string theory was a good mathematical model for nature and the universe at the smallest distance scales. (http://www.superstringtheory.com, 2010)

One of the most appealing aspects of string theory for scientists is that it nicely does what no other theory has been able to do in science so far. When we discuss gravity based on the super string theory, we are able to derive the theory of General Relativity without the insane correction factors that have so far haunted scientists. Another appealing aspect (for some) of string theory is that for this theory to be valid there are 10 or 11 dimensions that are currently hidden to us. So many interesting questions and theories can come out of this. For example, why can't we see these dimensions? And what kind of effect could understanding these hidden dimensions have on the world? These and other aspects of string theory have kept many scientists very optimistic of its possibilities.

Origins of Time

*"I never think of the future. It
comes soon enough."
Albert Einstein*

I think that time is one of the most familiar and most commonly used scientific concepts for most people. It's a dimension that we discuss in everyday life, but we rarely give it any deep thought. For most people, an hour simply consists of a certain number of minutes, a day of hours and a year of days. But we rarely think about the fundamental nature of time.

This chapter focuses on time; specifically it focuses on what it is, where it's from, how (if at all possible) to manipulate it and how we can better understand it. The new ideas that I'm going to present in this chapter may seem insane at first, but give them a little *time* to settle in and hopefully by the end they'll make a little sense.

In previous sections of this book, I've hinted at the possibility that quantum mechanics may not be able to describe both the complete nature of the subatomic world and of the macroscopic universe. I think most scientists would agree that we still have much to learn about the world around us. *Could it be possible that quantum mechanics is simply part of a complete solution that will hopefully someday lead to a better understanding to the subatomic world and to our universe?* I think that this is at least a possibility and by approaching quantum mechanics from this standpoint it may help to explain mass and time, and it will hopefully open some previously shut doors.

In order to really understand time and the universe, an entirely new theory may be necessary. This new theory will be as drastic of a change as the differences between Ptolemy's universe and Einstein's picture of the universe. This somewhat nonstandard approach to unknown problems is an approach that I believe would make even Einstein comfortable. One of Einstein's many frustrations with the scientific community of his time was their incremental approaches when attempting to solve problems. For example:

When incorporating gravity into special relativity, Einstein believed that an entirely new theory may be necessary, hence his Theory of General Relativity. Einstein believed that by simply adding to our current knowledge and by adding to his existing Special Theory of Relativity it would likely lead us down some dead end roads. Thus, he sometimes found it necessary to start from scratch.

I want to step back for a moment and recap where we are today regarding quantum mechanics. The more I read about the subject of quantum mechanics, the more humbled I become. I believe that we really don't know as much about quantum mechanics as some people may think. To help you get a sense of our lack of understanding about quantum mechanics, I'd like you to consider another seemingly pointless example:

Let's say an observer is watching a game of chess but he doesn't understand any of the rules about how to play. Let's also say that the observer had some bad food the night before, which makes him have to go to the bathroom every 3 minutes. During his frequent bathroom breaks, the game of chess continues. This competitive game of chess lasts about 4 hours (during which time the observer made dozens of trips to the latrine). If we were to ask the observer with the bathroom troubles after the game to explain the rules of chess, he may get a few of them correct. However, he would probably not know everything about the game due to his frequent absences. It would probably be a good assumption that he wouldn't be ready to enter a tournament. Later on, if this observer watched another game, he would be able to make some predictions about the rules but probably with a fairly low accuracy. More likely than not, his picture about the rules of chess would be far from complete.

How does this chess game example relate to quantum mechanics and time? Well, the connection is simple. Perhaps it's possible that the rules of quantum mechanics allow us to be able to predict results based on probabilities on a large scale, that is, the scale we can observe. But if we could observe a much smaller scale, say 10^{-34} m (sometimes called the Planck scale or Planck length, see Figure 17) perhaps we could accurately predict everything without any uncertainties or probabilities.

In order to make these accurate predictions, the first thing we would need to do is to understand what the universe looks like at this tiny subatomic scale. *At the heart of understanding this picture will be complete and accurate descriptions of space, time, and mass.* These are the three variables that are fundamental to our knowledge. Our ultimate goal is to understand them.

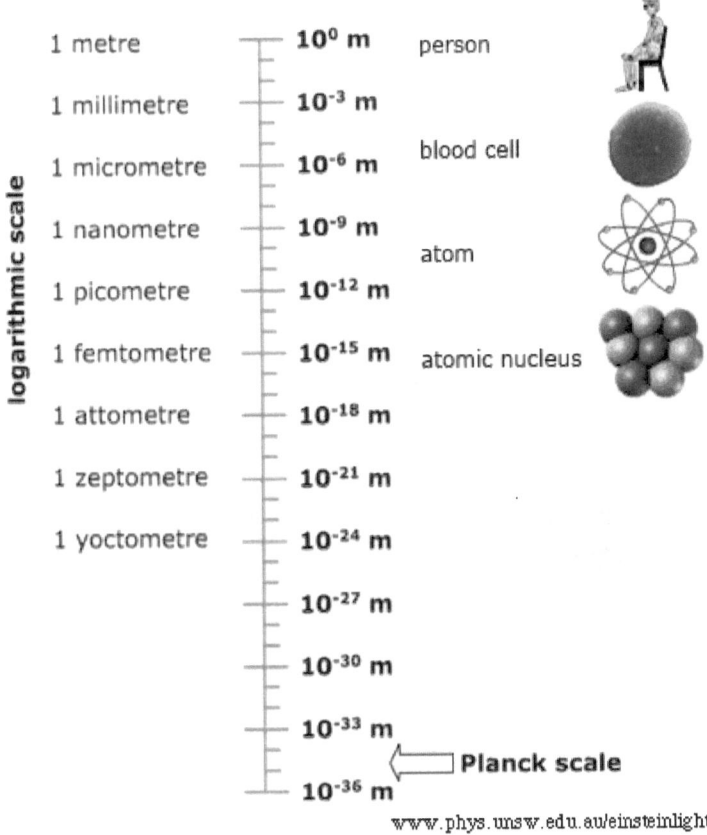

logarithmic scale

1 metre	10^0 m	person
1 millimetre	10^{-3} m	
1 micrometre	10^{-6} m	blood cell
1 nanometre	10^{-9} m	atom
1 picometre	10^{-12} m	
1 femtometre	10^{-15} m	atomic nucleus
1 attometre	10^{-18} m	
1 zeptometre	10^{-21} m	
1 yoctometre	10^{-24} m	
	10^{-27} m	
	10^{-30} m	
	10^{-33} m	
	10^{-36} m	Planck scale

www.phys.unsw.edu.au/einsteinlight

Figure 17: Planck Scale Comparison

To attempt to set the stage for the complete understanding of space, time and mass, I'd like to start at the beginning moments of our universe: *the Big Bang*. For this book, I'll assume that the Big Bang theory is completely correct, despite many thought provoking, competing theories.

Personally, I don't think I'll ever be able to fully accept the Big Bang theory for a few reasons. The biggest of which is my lack of understanding about what happened or what the universe was like before the "bang." You see, the Big Bang theory does not provide any explanation for an initial condition; instead, it describes in considerable detail, the general evolution of the universe since that time (although general relativity predicts a gravitational singularity before the bang). Putting this aside, I'll continue with a brief explanation of the Big Bang theory.

The Big Bang Theory

The Big Bang theory generally refers to the idea that the universe has expanded from a primordial hot and dense initial condition at some finite time in the past and continues to expand today. The earliest phases of the Big Bang are still mostly guesswork. However, in an article published by Steven Mostyn on Nov 9, 2010 titled, *"End of the world? LHC successfully creates mini-Big Bang"*, Mostyn describes that by smashing together lead ions instead of protons, researchers at the Large Hadron Collider have produced a "mini-Big Bang".

According to researchers, the collisions obtained at the Large Hadron Collider were able to generate the highest temperatures and densities ever produced in an experiment. Scientists have insisted that this process took place in a safe, controlled environment, generating incredibly hot and dense sub-atomic fireballs with temperatures of over ten trillion degrees, a million times hotter than the center of the Sun.

It's believed that at these temperatures even protons and neutrons, which make up the nuclei of atoms, melt resulting in a hot dense soup of quarks and gluons known as "quark-gluon plasma". As previously described, quarks and gluons are sub-atomic particles which are believed to be some of the building blocks of matter. In the state known as quark-gluon plasma, they are freed of their attraction to one another. This plasma is believed to have existed just after the Big Bang.

There's still much more research to be conducted at the Large Hadron Collider. But some of data released so far may be of use to us in this book which we'll come back to from time to time. Now, I'd like to describe the most common model used to explain the Big Bang in a little more detail with a fairly rough timeline to help put it in perspective.

Time = 0 to 10^{-37} Seconds after Bang →: In the most current Big Bang model, the universe was believed to be filled homogeneously and isotropically with an incredibly high energy density, huge temperatures and pressures, and it was very rapidly expanding and cooling. It's believed that at approximately 10^{-37} seconds into the expansion, a phase transition caused a cosmic inflation, during which the universe grew exponentially. (Guth, 1998)

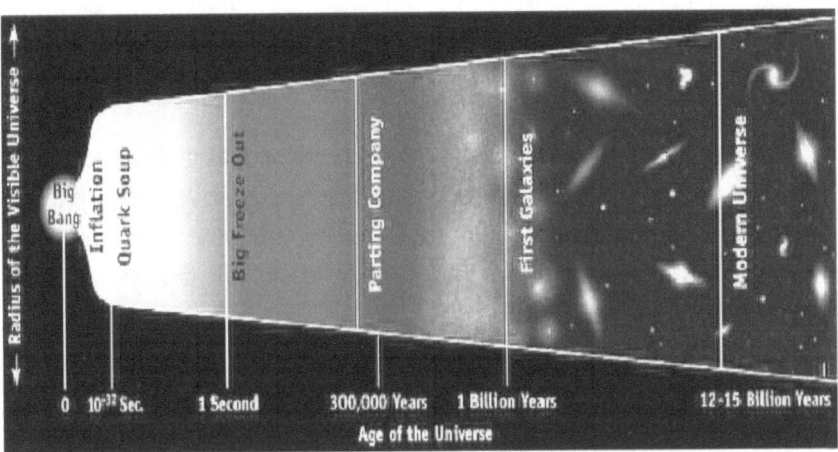

Figure 18: History of the Universe (http://cosmology.berkeley.edu, 2010)

Time = 10^{-37} to 10^{-11} Seconds after Bang → After the inflation stopped, it's believed that the universe consisted of quark-gluon plasma, as well as all other elementary particles, similar to what was observed in the initial LHC experiments. (Schewe and Stein, 2005)

Temperatures were so high that the random motions of particles were at relativistic speeds. Particle-antiparticle pairs of all kinds were being continuously created and destroyed in numerous collisions. At some point during the process, an unknown reaction called "baryogenesis" violated the conservation of baryon number, leading to a very small excess of quarks and leptons over anti-quarks and anti-leptons. The range of this disparity was of the order of 1 part in 30 million. It's believed that this resulted in the predominance of matter over antimatter in the present universe. (Kolb and Turner, 1988)

The universe continued to grow in size while the temperature decreased. This process caused the typical energy of each particle to decrease over time. Symmetry breaking phase transitions put the fundamental forces of physics and the parameters of elementary particles into their present form. (Kolb and Turner, 1988)

Time=10^{-11} Seconds to 10^{-6} Seconds after Bang ➔ After about 10^{-11} seconds, the picture becomes less speculative, since particle energies dropped to values that can be attained in modern day particle physics experiments (unlike most of the experiments taking place at the LHC).

Time=10^{-6} Seconds to Minutes after Bang ➔ At about 10^{-6} seconds, quarks and gluons combined to form baryons, such as protons and neutrons. The small excess of quarks over anti-quarks led to a small excess of baryons over anti-baryons (an anti-baryon is simply an antiparticle of the baryon, such as an antineutron or antiproton).

The temperature was now no longer high enough to create new proton-antiproton pairs (similarly for neutrons-antineutrons), so a mass annihilation immediately followed, leaving just one in 10^{10} of the original protons and neutrons, and none of their antiparticles. A similar process happened at about 1 second for electrons and positrons. After these annihilations occurred, the remaining protons, neutrons and electrons were no longer moving at relativistic speeds and the energy density of the universe was now dominated by photons (with a very minor contribution from neutrinos).

Time=Few Minutes to Several Thousand Years after Bang →
A few minutes into the expansion, when the temperature was about a billion Kelvin and the density was about that of air, neutrons combined with protons to form the universe's deuterium and helium nuclei in a process that's referred to as "Big Bang Nucleosynthesis" (Big Bang Nucleosynthesis is often abbreviated BBN and it generally refers to the production of nuclei other than those of H-1 during the early phases of the universe). (Kolb and Turner, 1988)

Time=Several Thousand Years after Bang Years to Present →
Most protons remained uncombined as hydrogen nuclei. As the universe cooled, the rest mass energy density of matter came to gravitationally dominate that of the photon radiation. After about 379,000 years, the electrons and nuclei combined into atoms (mostly hydrogen); hence the radiation decoupled from matter and continued through space largely unimpeded. This relic radiation is known as the cosmic microwave background radiation. (Peacock, 1999)

The Hubble Ultra Deep Field showcases galaxies from an ancient era when the universe was younger, denser, and warmer according to the Big Bang theory. Looking through the Hubble is like looking back in time. Over a long period of time, the slightly denser regions of the matter gravitationally attracted nearby matter and thus grew even denser, forming gas clouds, stars, galaxies, and the other astronomical structures observable today. The details of this process depend on the amount and type of matter in the universe (which we will discuss later in the book).

The three possible types of matter are known as "cold dark matter", "hot dark matter" and "baryonic matter". The best measurements available (from WMAP) show that the dominant form of matter in the universe is cold dark matter. The other two types of matter make up less than 18% of the matter in the universe. This will be discussed in much greater detail later in this book. (Hinshaw, G., et al, 2008)

New Ideas about "After the Bang"

For the rest of this chapter, I'll focus on the moments shortly after the Bang allegedly took place. I'll make some assumptions that will carry over for the rest of the book. I think that most of these assumptions and ideas represent a new way to picture the universe and a new way to explain some of the mysterious subatomic properties.

I'd like to first consider how the fabric of space and time looked shortly after the Bang. An important point, common to most theories, is that energy was converted to mass while the universe was expanding. Let's assume that to be true and also assume that during this entire process, the sum of energy and mass was conserved. That's basically the Conservation of Energy applied to mass as well (which in most texts is the Conservation of Energy). I think those assumptions are fairly safe but let's state them like this: *During expansion, mass was conserved, which includes the mass equivalent of energy. Essentially, that's Einstein's $E=mc^2$ formula.* Not much explaining should be required here.

I would also like to consider that during the expansion process the combination of space and time was conserved. This is a new idea that will require much more explaining. Let's state this new idea like this: *During expansion, space was conserved, which includes the Space Equivalent of Time.*

Before I explain what *"Space Equivalent of Time"* means, I want to briefly explain my motivation for grouping these properties this way. In my line of thinking, in order to fully rationalize the universe, I believe it may be necessary for us to consider extra dimensions or perhaps it may be necessary to ignore some dimensions. So what dimension could we possible ignore (or group with another dimension)? Answer: *Time*

Don't jump off the wagon just yet; please let me explain. I think it would be a big leap to say that we can ignore a dimension like time, but if we could, some pieces of the universe may fall into place. But we're not going to actually ignore time; we're just going to consider it as a function of space. Let me explain this a little more.

Time, by its very nature, is a difficult quantity to define and understand. Let's consider the ideas of Gassendi, a French philosopher, priest, scientist, astronomer, and mathematician. He was a truly gifted man who very few know anything about. He described the flowing of time by analogy with a river. This analogy raises an immediate problem: You see, flow presupposes speed of flow. We can say, for instance, that the water in the river flows at a speed of two kilometers per hour with respect to the bank as our standard of rest. But if his analogy is true, what speed can we assign to the "flow" of time, and with respect to what? What sense does it make to say that time flows at the rate of 1 minute every 60 seconds? The river could have been flowing at four kilometers per hour; but if the conditions were different, could time have been flowing at 2 minutes every 60 seconds? The difficulty seems to stem not just from the river analogy, but from the spatialization of time as such. Yet once we quantify time, the spatial analogy seems hard to avoid.

Meyerson wrote in Capek a very profound statement that's worth reading a few times., *"If becoming is to be transformed into being, so that the act of occurring becomes a simple unimportant formality for an event; if succession is only an illusion, and if every physical system constitutes a changeless whole - that can mean only one thing: the abolition or disappearance of time."* (Meyerson, in Capek, p. 355)

Another problem arises when we attempt to quantify time. Time as we experience is unidirectional, that is, it only goes one way. As soon as we quantify time, however, we have the immediate problem that any mathematical quantity can in principle either increase or decrease. That is, mathematical dimensions are "isotropic" by which I mean they're the same in both directions. Quantitative dimensions, especially when graphed, are governed by what Bertrand Russell (a brilliant British philosopher, logician, mathematician, historian, socialist, pacifist and social critic) called the "Axiom of Free Mobility:" We can in principle move either way along them. (Meyerson, in Capek p. 356.)

Mathematical time, therefore, automatically abstracts from one of the most essential features of time; it fails to capture the unidirectional nature of experienced time. In so far as we conceive or imagine time as a spatial dimension, we can dream about the possibility of moving along the time axis either way.

Many science fiction stories about time travel rely on our ability to treat the dimension of time as if it were a spatial dimension and our ability to travel in both directions. We'll discuss time travel in some detail later in this book.

For Newton, time was a pure, one dimensional quantity to be measured by clocks and other physical systems. Such measurements, however, always fall short of the ideal, because any human measuring system is to some extent irregular. Newton therefore distinguishes between time in itself and time as it is measured by us.

Absolute, true and mathematical time, of itself, and from its own nature, flows equably without relation to anything external, and by another name is called "duration". Relative, apparent, and common time, is some sensible and external (whether accurate or unequable) measure of duration by means of motion, which is commonly used instead of true time; such as an hour, a day, a month, a year. (Newton, Capek)

Absolute time, for Newton, is somewhat like a Platonic universal; it's time as it would be for God, who *"is everywhere present; and by existing always and everywhere, he constitutes duration and space....God is the same God, always and everywhere. He is omnipresent not virtually only, but also substantially."* (Newton, Capek)

Elsewhere, Newton refers to space and time as God's "sensorium". The being of time in itself -- that is, for God -- does not depend on our measure of it. God, then, by being omnipresent serves the function of an absolutely regular, universally intersubjective, clock, that is, one whose intervals are always of the same duration at every time and in every place, and for each subject. But could God impose irregular intervals to time as well?

As much as I hate to change courses again, it's necessary at this stage to briefly pause and discuss the three concepts of time we'll need to juggle:

1. Subjective, non-homogeneous time.

2. Relative (physical) time that human physicists measure by natural mechanisms;

3. Absolute (physical) time, measurable only by God.

It was Einstein who excluded God from the Cosmic Association of Physicists and permitted membership only to humans. This is not to say that he did not believe in God, which is what many people who study Einstein are taught. By doing this, the distinction between absolute and relative time then becomes impossible, since all time becomes a quantity measured by a human observer. In reality, then, all time measurements are relative to actual observers located at particular points in space and time, having their own clocks and moving at specific velocities and accelerations. This, more or less, is the Special Theory of Relativity as we discussed previously in Chapter 4.

"Scientists were rated as great heretics by the church, but they were truly religious men because of their faith in the orderliness of the universe."
- Albert Einstein

Simultaneity

Now I'd like to deal with the problem of simultaneity and later show the implications of this discussion on the flow of time. Eventually we'll show how it relates to Einstein's notion of the union of space-time.

I think we can all agree that *if* time is a quantity, then it must be measurable. To measure an interval of time is to compare it with some other standard interval (e.g., five seconds on a calibrated stopwatch). By comparing time, I mean

"We know from science that nothing in the universe exists as an isolated or independent entity."
-Margaret J. Wheatley

one must ensure that both events start at the same time and then one must observe whether they stop at the same time. So the nature of time as measurable is based on the notion of "at the same time" or "Simultaneity."

As long as we only have one observer, that is, as long as the intervals to be compared are in the same place, it's easy for the consciousness of the observer to judge the events as being simultaneous. But what does it mean to say that two spatially separated events occur "at the same instant" or are simultaneous? Here in lies the difficulty.

If there were a static substance (say for example a stationary time substance) spread over the entire universe so that the substance was present in all places, then simultaneity would be absolute for it would refer to the two events being present together in the unity of the one, non-dispersed, consciousness. As finite human beings, we might be able to simulate this effect if two observers in different places had a system of communication which was instantaneous. Then, as soon as an event was observed by observer A, that observer could send a message at an infinite speed to observer B who, observing another event in his or her vicinity at the same moment as receiving the message, could declare with great certainty that the two events to be simultaneous. Unfortunately an experiment like this is not so practical; in fact it's impossible, in real life.

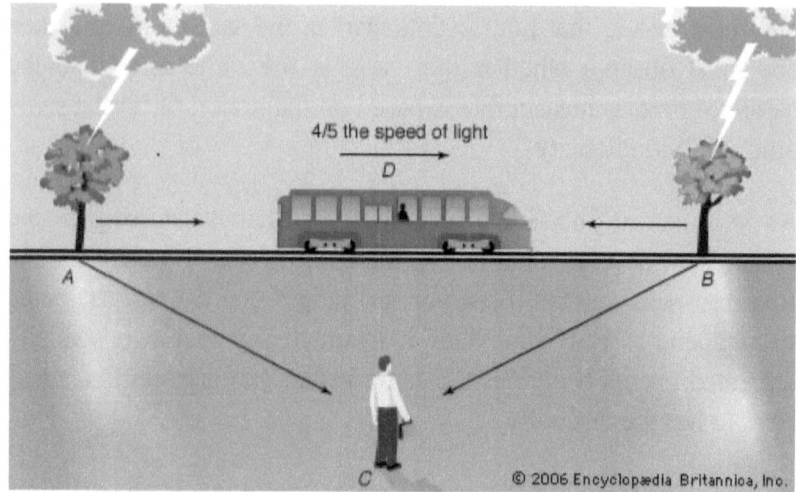

Figure 19: Simultaneity (Encyclopedia Britannica, Inc., 2006)

Figure 19 clearly shows, after some brief examination, the dilemma we face when dealing with simultaneity. In Figure 19, the observer at point C will observe the lightning strikes at points A and B and declare the events to be simultaneous. However, the passenger on the train, who is traveling very fast, will not observe the events at the same time.

Einstein hypothesized that there is no communication system with infinite speed; the best we can do is rely on light, or other forms of electro-magnetic radiation, whose fastest speed, in a vacuum, is around 300,000 Km per second. Accordingly, the observers have to make allowance for the time lag in messages they send each other, and while this may not seem difficult, it is no easy task. Let's try an example:

Imagine I set out in a spacecraft to visit my girlfriend on Mars (we all know that women are from Mars and men are from Heaven) and as I leave home (on earth) I send a radio message to her home on Mars to tell her I'm on my way. Unfortunately, at "exactly the same moment," she decides to come and visit me on earth, and sends me a message to my home as she leaves. Let's say that these beautifully poetic messages take 15 minutes to arrive to the planet they are traveling to. For practical purposes the two spacecraft departures are "simultaneous" if neither of them gets the other's message before they leave their home. However, there is a little leeway here. Even if I left earth 14 minutes later, neither of us would still have received the message in time, that is, the departures would still be "simultaneous" according to how we just defined them. This creates what is known as a zone of "practical simultaneity" which for this interplanetary love example is around 15 minutes, as shown in Figure 20.

More technically, what we are defining is known as a "zone of causal exclusion". Since no causal process can travel faster than light, the zone of simultaneity is the time span within which nothing at point A can be either the cause or the effect of an event at point B. (Robb, 1914)

It's pointless to say that within a 15 minute window one moment is really simultaneous with my departure. As humans, this in practice just cannot be determined. From our relativistic point of view, the idea of an absolute simultaneity makes no sense; we must choose some instant within the zone at a distance place and define it to be simultaneous with the present instant here. As Dr. Robb, who published the Theory of Space and Time in 1914, put it, *"There is no identity of instants at different places at all. ... The present instant, properly speaking, does not extend beyond here."* (Robb, 1914)

To help explain this a little better, cosmic space-time diagrams, recognized by the presence of the light cone are used (shown in Figure 20). Going back in time, this cone expands in space as well. The cone encompasses all events that can be monitored because there is sufficient time for light coming from the event source to have traveled to an observer at a different location.

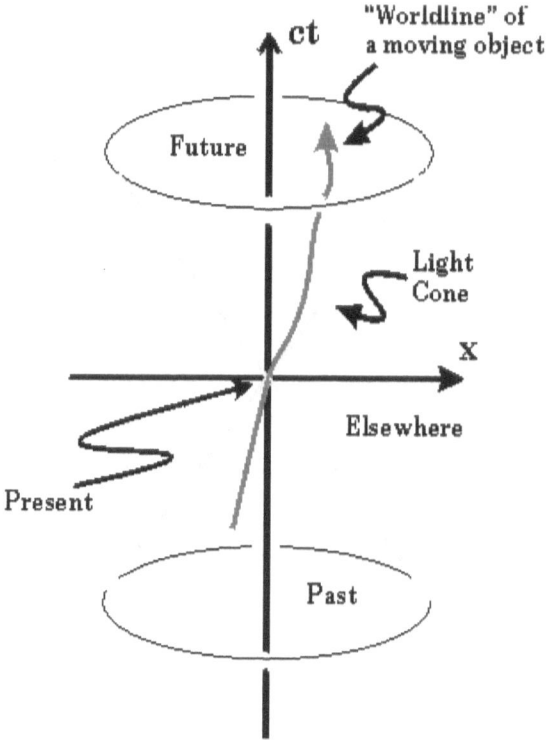

Figure 20: Cone of Space-time (http://rst.gsfc.nasa.gov, 2010)

The cone of space-time can be understood without suffering too much brain pain. A light cone is simply the path that a flash of light, emanating from a single event (localized to a single point in space and a single moment in time) and traveling in all directions, would take through space-time. To see that structure, we imagine an event at which there is an explosion. Light will propagate out from it in an expanding spherical shell. In two dimensional space, this will look like an expanding circle. Now we stack up these spatial snapshots (circles) to make a space-time diagram. The space-time diagram that corresponds to it will look like a cone.

This cone of space-time makes seeing into the past not as difficult as it may seem to be. All you have to do is look up at the stars on a clear night. The light from the stars has reached you (you can see the stars) so if the stars are millions of years away you could be seeing millions of years into the past. In fact, that star may have exploded thousands of years ago. This is because you are looking at the past light cone of those stars. A past light cone is everything that happened leading up to an event; a good way to describe this is to think of the sun burning out. If you looked up at the sun at the exact moment it burned out, you would see the sun as it normally is. You would be seeing everything leading up to its burning out - its past light cone. It would take eight minutes before the sun's future light cone reaches you, what the sun went through when it went out, the event itself and everything after it. This future light cone would then become the past light cone because the event already happened eight minutes ago.

Reichenbach, a leading philosopher of science, educator and proponent of logical empiricism who published <u>Axiomatization of the Theory of Relativity</u> in 1924 and many other scientific works, went on to state, *"The time-order of events separated by distance is arbitrary, within certain limits."*

How can we resolve this arbitrariness (not sure if that's even a word) and pick one instant within the zone as "simultaneous?" Let's look at another example to help explain this:

Let's go back to our example related to sending messages to/from my girlfriend on mars. Let's use that example and add that a UFO from a nearby galaxy is flying very rapidly over the solar system in the Earth to Mars direction (they just abducted some cows on earth and are heading home very fast). The UFO will pick up the message I sent from Earth but somewhat later (say five minutes later) than I might expect, since my message will have to chase the UFO to catch up with it. On the other hand, the UFO will bump head-on into my girlfriend's message from Mars and receive it earlier (again we'll say five minutes) than expected. The UFO captain will therefore declare that my friend left Mars ten minutes before I left Earth. Of course, another UFO speeding in the opposite, Mars-Earth, direction would say that the event of my departure occurred ten minutes before my friend's. That is, we can arbitrarily resolve the indefiniteness of the simultaneity by accepting to define simultaneity relative to one or both of these observers.

Additionally, there's also no use in saying that we can rely on the *"absolute"* judgment of an observer who is *"stopped"* somewhere between Earth and Mars. The solar system and our galaxy are themselves moving relative to other galaxies. The notion of being *"stopped"* cannot be defined. Like simultaneity, all velocities, being based on the measurement of time, are relative in the cosmos.

Therefore, the definition of which instants are simultaneous is relative to the velocity of the observer. The ramifications of this are far-reaching. It means that an event which occurs before another event for one observer may occur after it for another. The time-order of events is, within limits, not absolute, but dependent on the velocity of the observer. The arbitrariness of simultaneity means that the measure of any distant interval of time is arbitrary, since the two end-points are indefinite. Intervals of time, however, are part of the definition of velocity (distance/time), and velocity in turn is part of the definition of energy ($1/2\ mv^2$). Ultimately even mass, since its inertia is related to energy, must be seen as arbitrary, within limits.

The limits, of course, will depend on the speed of light. If light travelled slower, the limits would be wider; if light travelled faster, the limits would be narrower. Indeed, in the extreme case, if light were to travel at an infinite speed the arbitrariness would be reduced to zero and we would revert to Newton's absolute, God-based time. Since light travels at such a high a velocity, for most of the distances on Earth, and at everyday velocities, these limits are very narrow. Therefore, the world appears to us as it did to Newton. Relativistic phenomena only become apparent at much higher speeds and over longer distances.

I want to now look at some of the relativistic implications of time. These implications are very profound. If the length of time intervals depends on the velocity of the observer, then a period of, say, one hour as I observe it on my own watch may be measured as two hours by someone travelling very fast away from me, and as only a half hour to some approaching me. In other words, the *"flow of time"* may be faster or slower depending on who is observing it. We might expect that if two observers, A and B, are receding from each other and A observed B's clock to be running slow, than B would therefore think that A's clock was going fast. Surprisingly, this is not the case; each will observe the other's clock to be going too slow (as discussed in a previous example). If the velocity of recession was increased to close to the speed of light, both would see time in the other's neighborhood coming almost to a halt. (Remember my own observation of myself has no particular, Divine privilege; I too am speeding along at some velocity.)

How did Time Begin?

If we're going to have a chapter dedicated to time, we need to discuss and include an educated guess on how it began. This section will be just that, a guess as to how time began. First, let's consider the possibility that before the Big Bang time did not exist.

117

This assumption will be a problem for most people reading this book, but just humor me for a moment. Let's say that before the Big Bang, space existed as a finite point-source, or singularity, but moments later expanded quite rapidly. This agrees with our current Big Bang theories. *Out of this rapid stretching of space, time was created, or at least began to exist, as a dynamic measurable quantity.*

Many people have asked, "If the universe is expanding, what is it expanding into?" This seems like a fairly basic yet challenging question. Let's assume space itself stretched into the dimension of time. Time would therefore be its own dimension; it would be spread out everywhere in a non-uniform fashion. This sounds very similar to Newton's view of time. However, I'll expand on this and explain the differences.

Now I'd like to also assume that only after an interaction with mass, can time have any meaning or be observed. This would imply that time is a spatial dimension. But if time is a spatial dimension, how can we explain "the flow of time?" This has bothered me for quite some time. The questions and ideas in this section may seem pointless, but they are huge and they may help to connect and find some pieces to the giant puzzle of our mysterious universe.

One of the important questions that still remain is, "What was the universe like immediately before the bang?" This question is still open to much speculation. I'd like to present a possibility of what mass, time, and space were like before the Bang. Most scientists now believe that the universe is expanding and it is expanding at an increasing rate. If this is true, this fact may provide some important clues about the composition of our universe. At the moment just after the Big Bang, space and matter were created very rapidly. But where did this come from? Surely, matter could not come from nothing. I don't want to focus on the evolution of various particles, like the previous discussion on the Big Bang theory. I'm more concerned about the creation of mass and how time became a reality. These ideas will be far different than any ideas presented before.

Let's again look at the beginning moments of the universe. Now imagine that as space stretched, matter (mass) was created by the ripples in space. I'll go into this much further in the next chapter. For now, let's assume that time is simply the fabric over which space expands and mass is simply trapped energy confined by the ripples in space. But time, as we know it (based on Special and General Theories of Relativity), is not constant for all observers so it cannot be thought of like a piece of stretched fabric. *Time is more like a flux that travels through all objects with mass.* These objects with mass dent the fabric of space as shown in Figure 21.

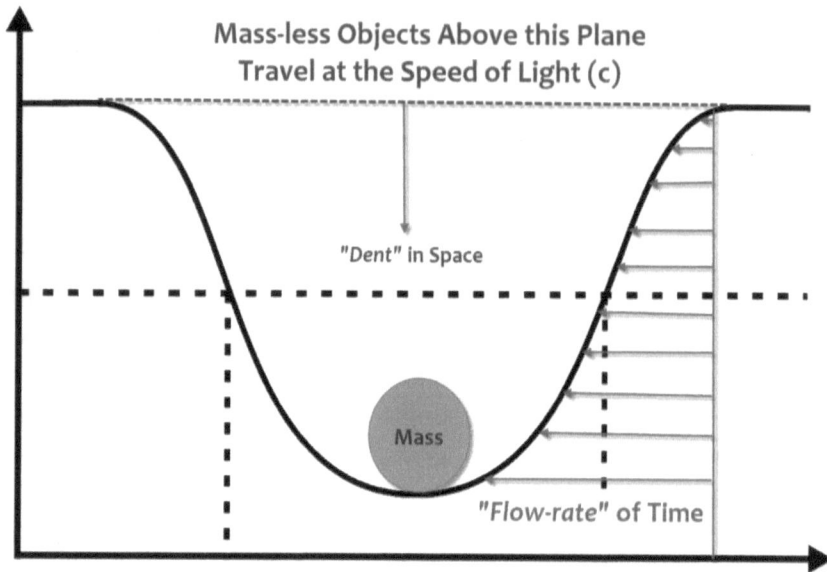

Figure 21: The Flow-rate of Time

Using Figure 21, when an object with mass accelerates, it begins to climb out of an imaginary potential well, shown above. The higher an object is in the well, the less the time flux passing through it will be. If an object could climb completely out of the well, it would cease to have any mass (no dent in space), and it would travel at the speed of light. In essence, it would be energy in its purest form.

In general, a flux is something (like energy) that passes through a certain area in a finite amount of time. For example, we may want to measure the amount of sunlight that hits a solar panel. The flux of sunlight is the amount of light that hits a given area over a certain period of time. If we reduce the angle of incident of the sunlight, less flux will hit the panel. The flux analogy for time may be useful to explain some of the physical properties that we observe.

Relativity, again...

Let's go back to Special Relativity for a moment. One feature we observe of Special Relativity is that time appears to pass differently in different reference frames. Let's say, for example, Observer A travels at .99c for a few hours, according to the clocks on his ship. When he returns and compares his clock to Observer B's clock, who was stationary relative to Observer A, the times will be drastically different. Observer B's clock may be several days ahead of Observer A's clock. How is this possible and how can we use Figure 21 (The Flow-rate of Time Diagram) to explain this?

If we consider time as a flux passing at the speed of light through all objects with mass, and we consider time as a function of space, we can begin to explain and understand this. When Observer A travels very fast, less time-flux will pass through him. Observer A is basically catching up to the time flux. This is a good thumb-rule but not very accurate. The real disparity between the clocks comes when the traveler jumps reference frames (changes directions or stops and reverses directions). During this period of changing velocities all the excitement happens.

How exactly does the flow of time change during this period of acceleration (positive or negative acceleration)? To explain this, I'll consider the simplest case: the spaceship slows down, stops, reverses directions, and then accelerates to the same magnitude of velocity as before but in the opposite direction. There will also be an observer at "rest" relative to the space traveler. This is the reference frame the clock was originally calibrated in.

As soon as the spaceship begins to slow down, the clock changes reference frames from the one it was initially calibrated in. At its original velocity, the clock was passing a certain distance over a given amount of time. This caused the flow of time to pass at a certain rate. During the deceleration, the space traveler passed over less time lines per unit of time than he did in the reference frame that the clock was calibrated in. Since less time lines were passed, his clock will run slower than it did in the original reference frame.

The idea of time-lines is one I haven't discussed yet, but it's useful in explaining this phenomenon. If we refer back to Figure 21 above (the Flow-rate of Time) we see that if an object is deeper in the space well, it will be exposed to more time flux. The idea of passing less time-lines is the same as rising up in the time potential well.

Now when the rocket speeds up in the opposite direction, the rate of passing timelines is still less than it was in the original reference frame. This causes time to pass at a slower, but at an increasing rate during the acceleration. Now, suppose the ship's velocity returns to the same magnitude, but in the opposite directions as it was during the 1st leg of the flight. Would the space-ship and the clock be back in the original reference frame?

My first guess at an answer to that question would be, *yes*. But this may not be, and likely is not, correct. The change in the flow of time only occurs when the two observers are travelling in different non-inertial reference frames. If two observers are travelling in opposite directions at the same speed, regardless of the speed, their clocks should read the same time. If one of the observer changes directions so as to travel in the same direction as the other observer, he has just changed reference frames (during the time when the change in velocity took place). Even though the space travelers will once again be traveling at the same velocity their clocks will not read the same time.

Earlier I mentioned that a common resolution to the twin paradox is that the traveling twin has to change directions to return home. This is what is most commonly discussed as the resolution but I think another, not so obvious, explanation also exists. When the twins were both on earth, their clocks were calibrated in the same reference frame. As soon as the twin took off in his spaceship his reference frame changed. You see, in order to take off from earth, a positive acceleration is required, causing a reference frame change. During this period, the traveler is in a non-inertial reference frame. At that very moment, time will pass differently for the two twins.

Length Contraction

I'll discuss length contraction in much more detail in the next chapter but I want to mention it here specifically in regards to the passing of time. According to Hendrik Lorentz, length contraction is the physical phenomenon of a decrease in length detected by an observer in objects that travel at any non-zero velocity relative to that observer. This contraction (more formally called Lorentz contraction or Lorentz–Fitzgerald contraction) is usually only noticeable, however, at a substantial fraction of the speed of light. Additionally, the contraction is only in the direction parallel to the direction in which the observed body is travelling. (Wikipedia, 2009)

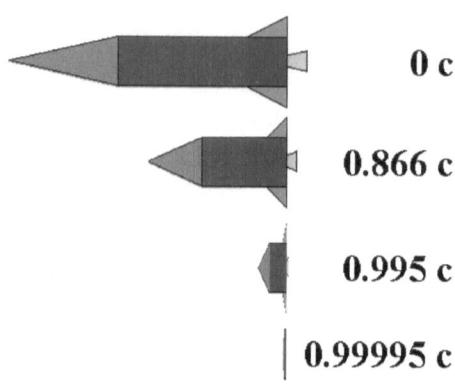

0 c

0.866 c

0.995 c

0.99995 c

Figure 22: Length Contraction (http://creationwiki.org, 2010)

For a moment, let's assume that length contraction and time dilation are intrinsically connected. They're kind of like love and marriage; you can't have one without the other (at least that's how it's supposed to be). As an object speeds, less time flux will travel through it. Less time-flux means that less space-lines appear to be crossed by Observer A. As a result, Observer A should return thinking he was gone for only a short time and traveled a short distance (relative to the stationary observer). This result is consistent with Special Relativity.

Now, what happens when Observer A, traveling at .99c, shines his headlights at Observer B who is stationary relative to observer A? This is a question similar to the one that Einstein asked, which I mentioned briefly in a previous chapter. I find the answer to this question to be quite surprising:

Answer: Both Observer A and Observer B will measure the same speed, the speed of light, that is: c. But how can they both be correct?

When Observer A shines the light, one might expect that he would measure a speed less than the speed of light, since he is also traveling very fast and in the same direction as the traveling light. But for Observer A, traveling at .99c, time appears to pass slower than it does for Observer B, since he is essentially catching up to the time flux. Therefore, as the light leaves Observer A, time "slows down" or passes slower that it does relative to Observer B. As a result, the light will appear to travel its distance over a shorter period of time. Therefore, the relative velocity between Observer A and the light will appear faster than we might expect, mainly c. This mental exercise can be explained using the idea of time dilation and length contraction.

Summary of Time

Finally it's time to summarize this very long, intense and at times quite boring chapter. Not a moment too soon for most. There were a few new ideas presented in this fairly long chapter. Some of these ideas are very significant, especially if proven to be true. In this chapter, *I surmised that time may have been created out of the original mass (or energy) from the Big Bang. This time acts like a flux that passes through all objects with mass. Since mass-less observables do not create a dent in space, they are not exposed to this time flux and therefore no time will appear to pass for them. It was the stretching of space that originally created the flux of time. Time and space, in this regard, are therefore interdependent properties.*

We also discussed simultaneity and the cone of space-time. Since simultaneity is an important consideration with time so too is location with space. Different observers may measure a different location depending on their speeds. Location can also be a relative property. This will be explored much more in the chapters to follow. Now, using this logic, one might conclude that if a rapid expansion or stretching of space created time, *then a rapid collapse or compression in space could destroy time.* But what exactly would destruction in time be like? And do these assumptions mean that space, time, mass and energy are all quantized properties? These questions are simply a brief preview to the chapters to follow.

The Nature of Mass

Chapter
6

"I'm fat, but I'm thin inside. ...There's a thin man inside every fat man."
George Orwell

This chapter, *the Nature of Mass*, will build on some of the new ideas that were presented in the previous chapters. It will also present some brand new ideas and theories. Eventually all this material will hopefully come together and make a little sense. Mass is a very complicated topic and it provided the original spark for me to start writing this book, so I want to spend a little time on it.

The exact nature and origins of mass have been puzzling scientists for decades as they still are to this day. And for good reasons; one of which is that scientists still no very little about mass. To start our exploration of this subject, I want to go back and try to understand the beginning of our universe, basically just the first few moments after the big bang. In the previous chapters, we briefly examined some of the popular theories about the origins of mass. I also proposed a new conservation of mass/energy and space/time theory. However, a couple questions still remain: *"How exactly is energy converted into mass?"* and *"What is mass?"*

It's helpful at this point to go back to Democritus' question about cutting cheese into very small pieces. Basically, Democritus asked what seemed like a simple question, "Can we keep cutting and cutting objects, specifically cheese, forever and still have cheese?" Unknowingly, he was asking, at the most fundamental level, what would a single quantity or quanta of cheese look like? Today scientists would refer to a single quantity of cheese as a cheese molecule. But rather than slicing cheese, I would like to ask the question," What would the most basic fundamental unit of mass look like?" I earlier stated, in one of my assumptions, that I believe that there's a point where mass cannot be divided any further. Whether or not this assumption is true may not seem terribly important, but it is. *It's possible that this one important assumption may hold the key to our entire universe.*

Scientists frequently discuss quantized energy states as if the idea of them is nothing more than common sense. The notion of quantized energy has been engrained in some of us for so long that it's hard to believe that anyone could have ever thought differently. But they did, however; in fact for many years.

It's also worth noting that the way most of us envision mass has changed little for most people for several years as well. It therefore seems like this subject presents a prime opportunity to introduce a new idea regarding mass for us all to consider. The new idea is not at all difficult to comprehend. Quite simply put:

Is it possible that mass too is quantized? Could the idea of quantized mass be feasible?

The idea of quantized mass is not one that I've ever heard discussed or read in any text. And at first glance it appears to make sense (at least it makes sense to me). You see if certain energies are quantized, and energy has some mass equivalence then mass too should be quantized, right? That seems like a rational argument. But I don't think we can make this giant conclusion so quickly. I believe this rationale is seriously flawed and I can assure you that it's not the sole basis for this new idea about quantized mass. But it does seem as though there are certain magic numbers that pertain to mass. For example, is the rest mass of all electrons the same? And if so, how can this be statistically possible? Surely this remarkable coincident defies simple logic and any rule of probability.

Another interesting and difficult concept to try to understand is why different particles have different masses. For example: Why does an electron have a different mass than a proton or a neutron? There must be a simple way to explain these interesting observations.

I'd like to now expand on the simple idea of quantized mass and later attempt to show how this idea can be explained. For now, let's assume that the mass of a particle is nothing more than packets of energy physically confined by the space around it as shown in Figure 23. In this picture, I've used a simple illustration of water waves to attempt to show how ripples in space may be able to trap energy and thereby create mass. *Could it be possible that when the right amount and correct type of energy comes close enough together, the space around the energy traps it and creates some type of subatomic particle which we perceive as mass? Our perception of mass may be nothing more than the interaction of light with the ripples in space.* We'll explore this idea much deeper later in this book.

Figure 23: How Energy Creates Mass

This new idea assumes that mass, at its most fundamental level, is nothing more than energy. This idea agrees with Einstein's mass-energy equivalence. An electron, for example, is not a particle orbiting a nucleus like what is taught in most elementary classes (I'm personally embarrassed that our public and private school systems still teach this concept). Nor is an electron simply a cloud of probability like we learn in more advanced physics classes. *But rather an electron is a bundle of energy confined by the space around it.*

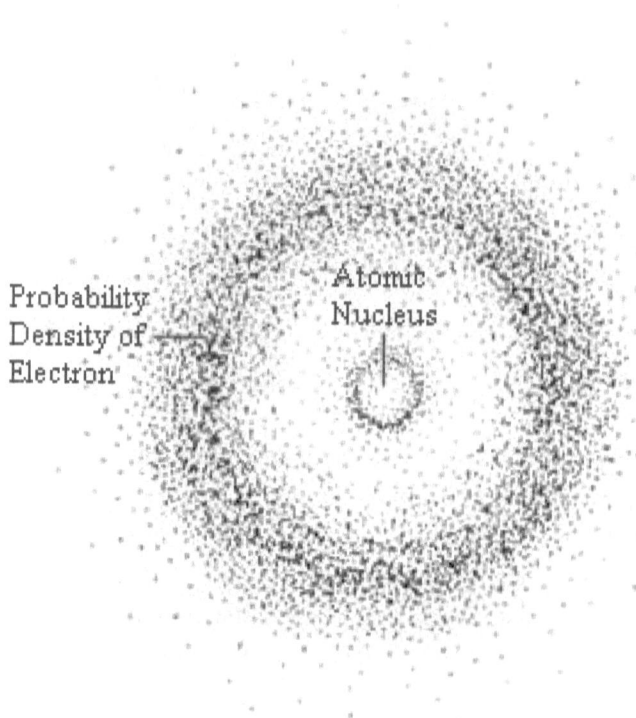

Figure 24: Electron Cloud (http://universe-review.ca/I15-53-quantum.jpg, 2010)

However, the process in which we observe an electron makes it appear to be a cloud of probability. So why then does an electron appear this way to us? One realistic possibility may be that we are simply limited by our technology and unfortunately we can't observe everything taking place on the subatomic level. Furthermore, in order to accurately describe the location of an electron we need a better understanding of the dynamic structure of the space surrounding the trapped energy, which we'll explore in Chapter 7.

The Higgs Particle

Scientists are feverishly hunting for, with some recent success, a very elusive particle that's believed to give mass to otherwise mass-less objects. This particle is known as the *"Higgs particle"*, the *"Higgs boson"*, or sometimes even the *"God particle"*. There's differing theories on what exactly a Higgs particle is. Some scientists believe it's possible that the Higgs particle could be a single particle, multiple particles, or even a manifestation of a field. Regardless, I believe my new trapped energy theory and the notion of a Higgs particle go quite well together. I also believe that this new idea may even help us to understand the nature of the Higgs particle, even if scientists are wrong. With that being said, I certainly don't believe that the hunt for the elusive Higgs particle is a lost cause. Let me explain why this is so important to science.

As previously mentioned, the Higgs particle is a predicted particle that's believed to give objects mass. The Large Hadron Collider (LHC) at CERN in Geneva, which is approximately 27 kilometers in circumference, is expected to provide experimental evidence either confirming or refuting the existence of the Higgs particle. By now (at the time you are reading this) the answer may already be known, but the information I present may still be useful for a deeper understanding.

Many discoveries and advances towards understanding the Higgs particle have already taken place at the LHC. While these experiments were taking place at CERN, similar experiments at the Fermilab Tevatron accelerator were ongoing focusing on attempts at detection (although the attempts are hindered by the lower energy capabilities of the Fermilab). It's been reported that the Fermilab physicists suggest that the odds of the Tevatron accelerator detecting the elusive Higgs boson are between 50% and 96%, depending on its precise mass (Morgan, 2009). The real promise for detection may be in the hands of the scientists at

CERN.

One direct consequence of the trapped energy theory is that all mass is created by energy being confined by the ripples in space and not by a particle itself. Depending on how this is interpreted, this could be in direct contradiction to the Higgs particle theory. But then again, it may not be, depending on how we choose to view the results. You see, this force mediating particle that we are searching for may actually be just the ripples in space that we "feel." *In other words, the Higgs particle itself may be the very structure of space that gives particles shape and mass.*

If a Higgs boson is detected, it could help to explain the origin of mass in the universe. More specifically, detection of the Higgs boson would explain the difference between the mass-less photon, which mediates electromagnetism, and the massive W and Z bosons, which mediate the weak force.

Higgs Field

The Higgs particle is believed to be one quantum component of the theoretical Higgs field. In empty space, the Higgs field has a non-zero amplitude, i.e., a non-zero vacuum value. The existence of this non-zero vacuum expectation plays a fundamental role in physics: it is believed that this field gives mass to every elementary particle which has mass, including the Higgs boson itself. In particular, the acquisition of a non-zero vacuum expectation value spontaneously breaks electroweak gauge symmetry, which scientists often refer to as the Higgs mechanism. This is the simplest mechanism capable of giving mass to the gauge bosons while remaining compatible with gauge theories. In essence, this field is analogous to a pool of molasses that "sticks" to the otherwise mass-less fundamental particles which travel through the field, converting them into particles with mass.

Everyday Analogy of the Higgs Field

I'd like to once again describe the Higgs particle and the Higgs field in a different, slightly simpler way that relates to the trapped energy theory, just in case it didn't sink in previously. Let's imagine that you're a famous rock star trying to make your way from the door of your favorite restaurant to your parked stretch limousine 150 feet away (I was going to use a famous scientists but I don't think it would be as convincing since few people have any admiration for them). As you make your way to the car, you're swamped by hundreds of reporters and fans. This mob of people will of course slow down your travel, creating what seems like a sort of viscous effect. As you keep walking, new people file in as you leave the others behind you. You finally make it to your limo, but you were slowed down considerably by the crowd. In this example, the reporters and fans were obviously the Higgs particles which in essence gave mass to the moving rock star, you.

The trapped energy theory can also be used to explain mass. As described earlier, objects with mass may create a dent in the space around them. This dent in space is what confines the energy. The dent itself acts like a potential barrier that impedes the motion of the trapped energy. In turn, we see and measure this as an object with mass. The larger the dent, the more massive the object and the more energy the dent can hold. It doesn't seem logical that a finite area in space could store an infinite amount of energy. It's limited by the volume around it. In the next chapter, we'll discuss rules on how this energy can be stored.

I've given some consideration that the dent in space is the energy mechanism we observe. In other words, I've considered that there is no energy confined by space; it is the fabric of space that we perceive as energy. When the space is collapsed, a wave propagates and we measure this as some form of energy. It's an important difference than space confining energy. Rather it is the space itself that is energy. This is something that needs further discussion. But for now it's just another theory of mass for us to consider.

Atom Smashers May Have Found a New Force?

This was an interesting article I came across that was posted in 2011 by Fabrice Coffrini of AFP News titled, _US Atom Smashers may have found a new Force_. The article discusses some of the differences of opinions scientists have with the Higgs Field and possible alternatives for this new force. The following is the meat of the article:

Physicists will announce Wednesday that data from a major US atom smasher lab may have revealed a new elementary particle, or potentially a new force of nature, one of the researchers told AFP.

The discovery is believed to relate to mass and how objects obtain it -- a persistent riddle to experts and one of the most sought-after answers in all of physics.

"There could be some new force beyond the force that we know," said Giovanni Punzi, a physicist with the international research team that is analyzing the data.

"If it is confirmed, it could point to a whole new world of interactions," he told AFP.

While much remains a mystery, one thing researchers agree on is that this is something beyond the "God Particle," or the Higgs-boson, a hypothetical elementary particle which has long eluded physicists who believe it could explain why objects have mass.

"The Higgs-boson is a piece that goes into the puzzle that we already have," said Punzi. "Whereas this is something that goes a *little bit beyond that -- a new interaction, a new force."*

For more than a year physicists have been studying what appears to be a "bump" in the data from the US Department of Energy's Fermi National Accelerator Laboratory, which operates the powerful particle accelerator Tevatron.

Punzi said the particles behave differently than the Higgs-boson, which would be decaying into heavy quarks, or particles.

The new discovery "is decaying in normal quarks," Punzi said.

"It has different features," he added.

"One thing we know for sure -- it is not the Higgs-boson. That is the only thing we know for sure."

Physicists were to discuss their findings further in a meeting to be webcast at 2100 GMT.

"Nobody knows what this is," Christopher Hill, a theorist at Fermilab who was not part of the team, was quoted as telling the New York Times.

"If it is real, it would be the most significant discovery in physics in half a century."

The Standard Model, again...

Unfortunately, we need to once again briefly discuss aspects of the standard model and describe how this model relates to mass. Since the Higgs field is a scalar field, the Higgs boson has no spin, hence no intrinsic angular momentum. The Higgs boson is also believed to be its own antiparticle.

The standard model does not predict the mass of the Higgs boson.

If the mass of the Higgs boson is between 115 and 180 GeV/c^2, then the standard model can be valid at energy scales all the way up to the Planck scale (10^{16} TeV). Many theorists expect new physics beyond the standard model to emerge at the TeV-scale, based on unsatisfactory properties of the standard model. The highest possible mass scale allowed for the Higgs boson (or some other electroweak symmetry breaking mechanism) is 1.4 TeV; beyond this point, the standard model becomes inconsistent without such a mechanism because unitarity is violated in certain scattering processes. Many models of super-symmetry predict that the lightest Higgs boson will have a mass only slightly above the current experimental limits, at around 120 GeV or less.

Mass-Energy Equivalence

In physics, mass–energy equivalence is the concept that the mass of a body is also a measure of its energy content. In this concept the total energy, E, of a body at rest is equal to the product of its rest mass, m, and a suitable conversion factor to transform from units of mass to units of energy. If the body is not stationary relative to the observer then accounts must be made for relativistic effects where m is given by the relativistic mass and E the relativistic energy of the body. Albert Einstein proposed the mass–energy equivalence in 1905 in one of his papers entitled, *"Does the inertia of a body depend upon its energy-content?"* (Einstein, 1905) The equivalence is described by Einstein's most famous equation:

$$E=mc^2$$

where E is energy, m is mass, and c is the speed of light in a vacuum.

The formula does not depend on any specific system of measurement units. For example, in natural units, the speed of light is set equal to 1, and the formula becomes the identity $E = m$; hence the term mass–energy equivalence. (Flores, 2010)

The equation $E = mc^2$ indicates that energy always exhibits mass in whatever form the energy takes. (Tipler, 2002) It does not imply that mass may be "converted" to energy. The modern theory holds that neither mass nor energy may be destroyed, but only moved from one location to another. In physics, mass must be differentiated from matter. In cases where matter particles are created or destroyed, the precursors and products retain both the original mass and energy, which is unchanged. Mass-energy equivalence also means that mass-conservation becomes a re-statement of the law of energy-conservation, which is the first law of thermodynamics (energy can be transformed but cannot be created or destroyed).

The equation, $E = mc^2$ has sometimes been used as a simplified explanation for the origin of energy in nuclear processes. But in reality, mass-energy equivalence does not explain the origin of any such energy. Instead, this relationship merely indicates that the large amounts of energy released in such reactions may have a large mass-loss. This mass-loss may be measured when the released energy (and its mass) has been removed from the system being measured. The equation, $E = mc^2$ also does not explain any of the processes by which the conversion of mass to energy takes place.

Once again, if we consider that mass is simply trapped energy confined by the space around it, we may be able to rationalize Einstein's famous equation, $E = mc^2$. It appears logical to conclude that when the conversion of mass to energy takes place, it's possible that space surrounding the energy is simply unwrapped. The unwrapping of the space removes the potential barrier around the energy and it allows the energy to be released and escaped. When scientists describe this process, they simply see it as the transformation of mass to energy. *But it could be possible that this process is just the unwrapping of space. To be consistent, energy has not been created nor destroyed in this process. The energy that was contained in the space around it has simply been released.*

Space Paths

*"Don't see things as they are,
and ask why. See things as they
could be, and ask, why not?"*
-Anonymous

his is the chapter where things will hopefully start coming together; the so-called chapter of enlightenment. So far we've discussed in some detail, time and mass. Much of what was discussed were new ideas and theories. Now we need to roll up these new ideas about time and mass with a discussion and new understanding of space.

The measurement of space is not easy and is subject to the same relativity as the measurement of time, and for the same reasons. As previously discussed, the length of a body as measured by a moving observer will be shorter than if the body is measured by an observer at rest relative to the body being measured. What is constant for all observers is the space-time distance between two events, that is, the length of the line joining the two event-points in a four-dimensional graph (technically, the square root of $x^2 + y^2 + z^2 + t^2$).

Observers at different velocities will attribute different components of the "distance" to the space coordinates and to the time coordinates. Somewhat loosely we can say that what is measured as a space distance to one observer will be measured as a time difference to another. In the case I mentioned above, when traveling at close to the speed of light, when time slows almost to a stop, the length of the observed moving object will be reduced almost to zero. As Einstein put it:

"Neither the point in time at which an event takes place nor the

point in space in which a thing takes place have any physical reality, but only the event itself, ... so that neither an absolute spatial relation nor an absolute temporal relation exists between two events, but only an absolute spatio-temporal relation. ... It is impossible to divide the four-dimensional continuum into a three-dimensional spatial continuum and a one-dimensional temporal continuum in any way that makes sense from the objective point of view. " (Einstein, Capek)

What does this statement imply about the spatialization of time? If, as Reichenbach argued, the quantification of time and its consequent spatialization do not change our conception of time which remains very different than space, then how can we say that time can become space and space time, or at least that they are interchangeable under certain observational conditions? (Einstein, Capek)

It's an interesting idea to think that time can become space and that space can become time (or at least that the two are interchangeable). But is this actually possible? For a moment, let's examine space as it pertains to quantum mechanics. Perhaps quantum mechanics does not directly predict the behavior of subatomic particles. *Perhaps quantum mechanics simply predicts the behavior of the space that the subatomic particles travel through.* In other words, we can observe particles and atoms and predict certain quantities, but maybe the particles are just obeying the rules of the space and the rules of time that confines them.

Could it be possible that the geometry and structure of space and time are actually fundamental properties? The exact way that space and time emerge may hold the solutions to many problems. If space and time are fundamental properties it could help to answer some important questions in physics. Let me give a few examples.

Fermat's Principle of Least Time

In most college level physics courses, *Fermat's Principle of Least Time* is discussed but usually not in great detail. Usually, this principle is only discussed to explain or derive other equations like *Snell's Law* (Snell's Law is not terribly relevant to this book so it won't be discussed here). But the *Principle of Least Time* may be much more significant than it first appears and it may be a clue to the nature and to the contours of space and our universe.

Simply put, the *Principle of Least Time* tells us that light will "choose" the path that takes the least amount of time for it to travel. If you think about that for a moment, you'll probably agree that this statement or principle is nothing short of astonishing. I remember listening to a lecture some time ago by the great physicist, Dr. Richard Feynman. During the lecture, he explained that light was somehow able to "sniff" out the path that would take the least amount of time to travel. That sniffing ability of light has always been difficult for me to comprehend and accept. It's been one of many unresolved issues for me and other scientists for several years.

You see, light does not gradually change directions, like a car does when it turns around a corner. But rather, light changes directions abruptly when it reaches the path of least time. In mathematical terms, the slope of this line would be undefined. So how is it possible that light "knows" which direction to take before it starts? How does it know which direction will take the least time? Does information somehow get transferred to the light while it's traveling? This cannot be possible. For information to travel back and forth soon enough to cause an action, the information would have to travel faster than the speed of light. We know from Special Relativity that information cannot travel faster than light.

So how could light "know" which course to choose? Something else must be happening here.

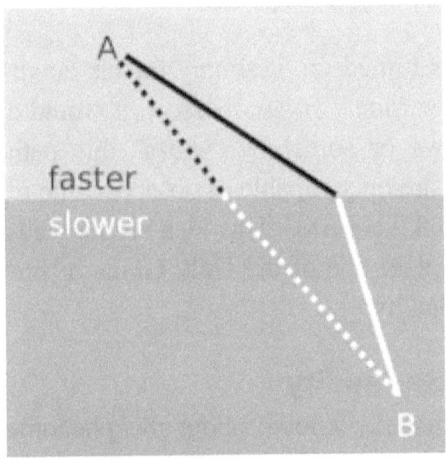

Figure 25: Fermat's Principle of Least Time (Crowell, 1998)

Figure 25 shows that light will not always travel in a straight line when going from an arbitrary point A to point B. In this figure, light is able to travel faster in the top medium than it can in the bottom medium. So rather than going down a straight line path from point A to point B (represented by the dotted line in Figure 25), light travels a longer distance in medium A and less of a distance in medium B (compared to the straight line path). This minimizes the overall time of travel between points A and B.

But how does light know to do this before we shine the light? Could it be possible that the light and other subatomic particles simply travel down the paths with the least amount of *spatial resistance*? Similar to my earlier example of crossing a field to reach a tree, in space there are seemingly an infinite number of ways to go from point A to point B. These "space paths" are everywhere and extend in every possible direction. Furthermore, they are constantly changing. *For light, the path of least spatial resistance is also the path that takes the least amount of time.* This path is etched into the fabric of space and time and account for the entire structure of the universe.

When a light is turned on, it simply travels down the path in the direction of least time. To an observer, it would appear that light somehow knows or somehow "sniffs" this path. Due to our limitations, we are only capable of observing the result and not the actual process taking place. This idea seems plausible and it may also help to explain one of the Holy Grails of physics, known as wave-particle duality.

Wave-particle Duality

Very little is actually known about the phenomenon known as wave-particle duality. Sure we can observe what it is and we can make predictions based on our theories but we don't really understand *why* this happens. Basically, the concept of wave-particle duality states that light can act either as a wave or a particle depending on our experiment and how we make our observation. The most common way to explain this is to examine *Young's Double Slit Experiment.* The double-slit experiment involves shining a light or a particle (atoms could be used instead of light) at a double slit, with the slits very closely separated (close relative to the wavelength of light). When light is shined, an interference pattern will form on a screen on the other side of the slit (see Figure 26).

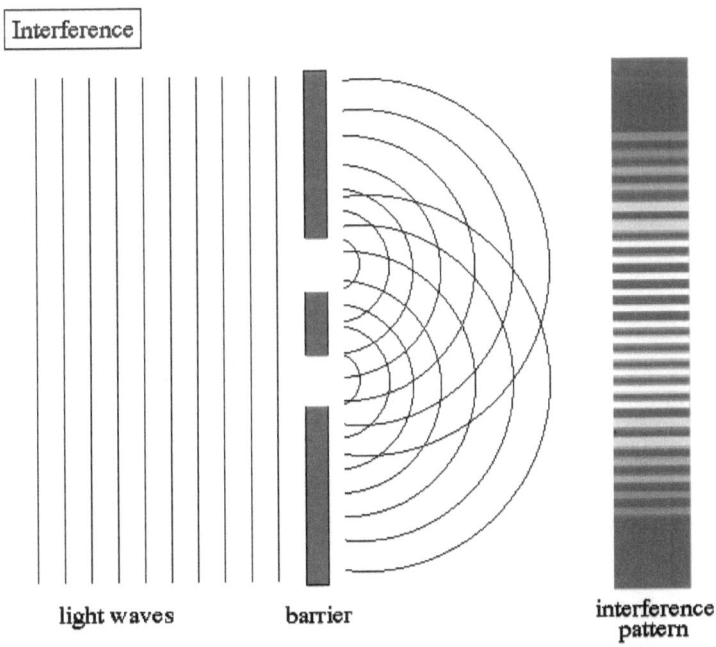

Figure 26: Young's Double Slit Experiment (http://abyss.uoregon.edu, 2010)

The results are consistent if electrons or atoms are used instead of a beam of light. When particles (such as electrons) are fired at the slits one at a time, over a period of time the electrons will hit the screen and a similar pattern will be observed as if the electrons were a wave of light.

Basically, the electrons will appear to form an interference pattern on the screen. But if only one electron was shot at a time, what exactly could it be interfering with? Is it somehow interfering with itself? The result of this experiment suggests that light and electrons (and other subatomic particles) can behave as a wave or a particle.

That's not the end to the surprising results from this experiment. If we try to determine which slit an electron or light travels through, using some sort of detector, the interference patterns will disappear. When this is done, the light or electrons appear to behave as particles, not waves. This interesting result can be viewed as the collapse of the wave-function. Basically, if a measurement is repeated without re-preparing the state (or conditions), one will find the same results as was found in the first measurement. When the quantum state is updated, the wave-function collapses.

The *Copenhagen Interpretation* describes this aspect of quantum mechanics. To put it simply, the Copenhagen Interpretation is a widely accepted interpretation of quantum mechanics. A key feature of quantum mechanics is that the state of every particle can be described by a wave-function. A wave-function is a mathematical representation used to calculate the probability for the wave to be found in a location or a state of motion. According to the Copenhagen Interpretation, the act of measurement causes the calculated set of probabilities to collapse to the value defined by the measurement. This feature of the mathematical representations is known as wave-function collapse.

There are other variations of this experiment that are equally perplexing that we really don't need to spend time discusing in this book. The main take-away from this section is that depending on how an experiment is conducted the light or particles may or may not behave like a wave.

For decades following the original double-slit experiment (in this experiment, they only performed and observed an interference pattern from light), Newton's *Particle Theory of Light* was dismissed. After the double-slit experiment, the scientific community was convinced that light was a wave. But then, at the beginning of the 20th century, along came Einstein who showed the world via the photoelectric effect that light could also have particle-like characteristics. The results later became known as wave-particle duality. Depending on the experiment, light could behave as a wave or particle. But how can this be possible? Can light really change its properties or could there be something else taking place that we are unaware of?

For now, let's consider the possibility that the double slit experiment does not explain the nature of light or the nature of particles involved but rather it explains the nature of the space around the light or particles. Is it possible that this is the same mechanism that we see with *Fermat's Principle of Least Time*? Perhaps, particles and light cannot choose how they behave, but instead are forced to travel in accordance with certain rules of space that we have yet to understand.

For a moment, let's look at how these ideas specifically relate and help us to explain the results we observe with the double slit experiment. I'd like to also try to explain how the contours and rules of space may determine how particles and light behave. But first, here are two important observations that we need to contend with:

1. Interference:

When two closely separated slits are open and a particle or light is shot at the slits an interference pattern will be formed on a screen behind the slits. This can even be observed when one atom is shot at a time. In the scenario where one atom is shot at a time, eventually, an interference pattern will develop. This has been used to show that particles undergo interference and behave like waves. But if they are shot one at a time, what are they interfering with? Does it really make sense that when a single particle is shot, it changes its form into a wave, then interferes with itself, then somehow recombines into a particle to make a single impression on the screen? It's difficult for me to accept that idea. An easier explanation could be that space-time is somehow shaped such as to cause the appearance of particle interference. Let me explain a little further.

If we use the idea of space contours (or space paths), then there are seemingly an infinite number of ways for an atom to travel from point A to point B. When a particle is shot, it simply travels towards the slit, down the space path that has the least space-time resistance. In a sense, the path lines of space-time are interfering with each other. The particles or light themselves do not interfere. *When particles or light travel down a space-time path, the potential for that path changes for the future light or particles that are shot at the slit. This makes that path less desirable for a period of time and forces a new route for the next particle.* The result is that the next particle will likely travel in a different space-time path to reach its endpoint.

This idea makes even more sense if we consider that the particles are nothing more than trapped energy, confined by the ripples in space. These space ripples interfere like the ripples of a pond. Due to this rippling, over time an observer will see an interference pattern develop.

2. Disappearing Interference:

If we place a detector on the other side of the double slit but before the screen, we get some interesting results. With the detector off, we observe a normal interference pattern. With the detector on, the particles hit the screen inline (more or less) with the slits and show no signs of interference. Some of the particles will spread out slightly, similar to a diffraction pattern, but there's no evidence that they interfere with each other. These results could be due to the effect of the measuring device. In the Copenhagen Interpretation this is viewed as the wave-function collapse. Basically, no matter how unobtrusive we try to construct our detectors, they will always cause some changes. The changes caused by the detector are not, however, in the particles but rather the changes are in the contours of space.

Now I would like to explain how we may be able to apply this new idea. The idea of confined energy to create mass may help to answer some other questions. For example, the speed of light pops up into many of the most fundamental scientific equations. For a brief moment, let's reexamine one of the most famous equations; an equation by Einstein: $E=mc^2$. This equation shows us among other things that there is an energy equivalent for mass. Basically, mass can be converted into energy and energy can be converted into mass.

To understand this a little better, we need to explore the basics of nuclear fission. During a nuclear fission reaction, a heavy unstable element breaks up (fissions) into two or more smaller more stable elements. The combined mass of the resultant elements is less than the original mass of the heavy unstable element. The difference in mass is converted into energy and this process is governed by Einstein's equation: $E=mc^2$ (See Figure 27).

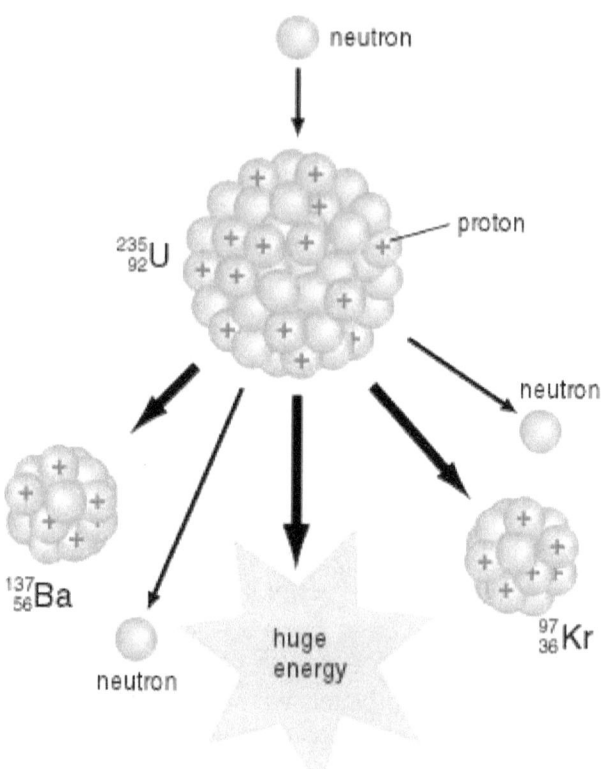

Figure 27: Nuclear Fission (http://www.kutl.kyushu-u.ac.jp, 2010)

While that may seem quite basic, there's still a lot to understand about mass and energy. Here are a couple of the lingering questions that were not answered in the mass-energy equivalence section of this book:

1. How does the speed of light factor into the equation?
2. Why can't we predict when and which atoms will fission?

As mentioned, the speed of light plays a role in many fundamental equations. We constantly see it pop up in a number of different formulas. In fission, it's thought of as a convenient conversion factor. But if we assume that particles with mass are nothing more than confined energy waves, we may be able to rationalize this. Here's how:

During the fission process, the space around some of the trapped energy waves is unwrapped. This causes the previously confined energy waves to be released, traveling at the speed of light. The released particles do not travel at the speed of light, just the energy with zero rest mass. The particles cannot travel at the speed of light because they cause a dent in space. The term c in the equation is squared because the light travels similar to a flux (it travels through a cross-sectional area) and therefore the quantity is squared. The loss in mass corresponds to the energy released from the vanishing particles.

Quantization

Now let's come back to the problem of understanding and describing mass and what that has to do with quantization. The idea of quantized properties for subatomic particle has puzzled scientists for the better part of a century. A couple of the troubling questions include:

- Why are only certain energy levels allowed for certain particles?

- Why are there forbidden regions where electrons can't travel?

When electrons travel between states, we basically observe that the electrons simply "jump" between states rather than smoothly traveling between the orbits (like most people would expect). There are locations where we do not observe electrons. These locations are known as forbidden regions. Currently, most scientists believe that this phenomenon is governed by probabilities.

For a moment, imagine a nucleus resting in space. The mass of the nucleus creates a small dent in the space around it. This dent in space causes a sort of ripple wave to occur moving away from the nucleus. The low points of the ripples in space create all of the allowed locations for storing other quantities of energy. This energy is what we observe as electrons.

Using this idea, we see that the electrons are not particles orbiting around the nucleus like most people are taught in school. But rather the electrons are energy waves spread out in the low potential locations in the space around the nucleus. Each one of these locations can store a certain amount of energy.

The rippling of space causes the forbidden regions that scientists observe. The electrons (or energy) are able to tunnel through the ripples in space rather than traveling over them. Thus the electrons appear to just show up in the new orbit rather than transiting smoothly between them.

Quantization of Light

Can any of these new ideas be used to explain the quantization of light? I believe they can, and here's why:

We know from the photoelectric effect that when a certain frequency of light, above some threshold, is shined on a metal an electron can be emitted. If the frequency is below the threshold then increasing the intensity of light (no matter how powerful the light) will not free any electrons. This is a fascinating observation that Einstein explained by theorizing that light came in discrete packets or bundles of energy, called photons. The energy of these photons can be formulized by: $E=h\nu$.

where h is Plank's constant and ν is the frequency of light

Believe it or not, it may also be possible to explain the

photoelectric effect using the space contour idea. Could it be possible that energy cannot be absorbed by the allowed locations unless the frequency of the energy coincides with the space pattern created by the surrounding masses?

In other words, the frequency of light must fit inside the spatial constraints of the space ripples. The light must be in phase with or above the frequency of the space contours. If the frequency of light is below the threshold it cannot be deposited in the space ripples. The energy of the light will, therefore not be deposited and it will not be able to free an electron. It's a new way to look at an old classic problem.

Could Faster than Light Speeds be Possible?

Referring back to Chapter 4, Einstein proposed that the speed of light was the universal speed limit for particles with mass. This idea opened the door for Special Relativity and gave us a new way to view the universe. But recent data from the CERN particle accelerator suggests that there may be a flaw in this idea. In September, 2011 data was released and collaborated on by 174 physicists that suggested that during the course of measuring the speed of over 16,000 neutrinos racing through their accelerator, that the speed required for them to complete the course was slightly faster than the speed of light (somewhere on the order of the particles arriving 17 millionth of a second early than a beam of light).

While this time difference may seem insignificant and knowing that the mass of a neutrino is very small it still poses a direct conflict to a very old and widely held belief that no particle with mass can travel faster than the speed of light. It therefore requires new ideas or better scrutiny of the scientific data because no matter how small this difference may seem it is anything but insignificant.

My theory suggests that maybe the particles are not traveling faster than light but rather the distance they are traveling may actually be shorter than believed due to the ripples in space.

SPACE PATHS

Dark Matter & Dark Energy

"I would draw a picture of it, but I can't."

I've included this brief section about dark energy and dark matter because of how little is known about these topics in the scientific community. I'm certainly not an expert on this subject, but I'll discuss what little I know. I've heard a recent estimate that stated about 75% of the energy in the universe is made of dark energy and 21% of the mass is made of dark matter. The remaining 4% is made up of the stuff we see every day.

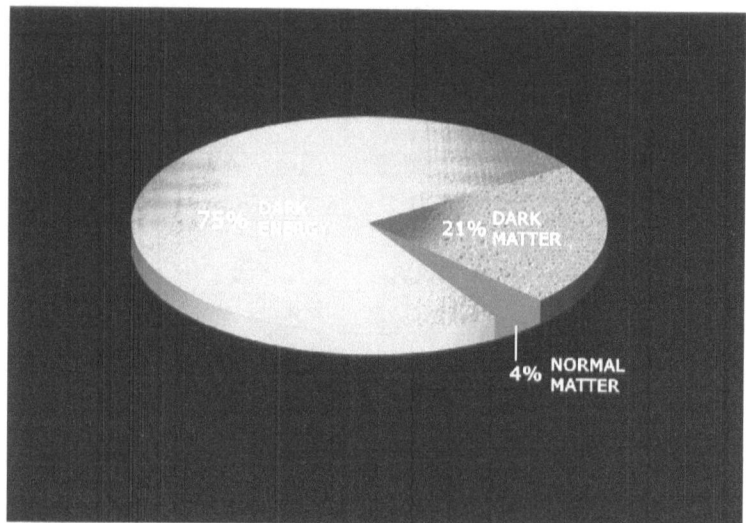

Figure 288: Distribution of Mass & Energy Throughout the Universe

The amount of dark energy and dark matter in the universe is so extraordinary that it certainly can't be ignored. If we actually understood what dark matter and dark energy were, it would help a great deal in understanding the universe. But the simple fact is that we don't. Some scientists believe that dark matter could be made up of photinos, the theorized super-partner of the neutrino. There are several other competing theories among scientists, most of which we won't explore in this book.

If you're like me, you're probably wondering, "How exactly did scientists come up with the numbers regarding the percentages of the universe believed to be composed of dark energy and dark matter?" The short answer is that I'm not really sure. I believe that it had something to do with the way galaxies spin. A number of years ago scientists observed that galaxies were spinning faster than they expected. Because of this rapid rotation, scientists believed that there was much more mass out there than what they could actually see. This missing mass, like a sea of particles spread out unevenly through the universe, became known as dark matter.

Even though we don't know very much about dark matter and dark energy, I trust that the scientific estimates are fairly accurate. These estimates are, after all, based on very comprehensive observations taken over many years by several different scientists.

Now I'll give you a little background on dark energy and dark matter and how scientists believe they came into existence. Please keep in mind that I'm not an expert on this subject, but the information presented in this chapter was developed by experts. When the universe was young, it was believed to be smooth and featureless. As it grew older and developed, it became organized. We know that our solar system is organized into planets orbiting around the Sun. On a scale much larger than the solar system (about 100 million times larger), stars collect themselves into galaxies. Our Sun is an average star in an average galaxy called the Milky Way. The Milky Way contains about 100 billion stars. On still larger scales, individual galaxies are concentrated into groups, or what astronomers call clusters of galaxies.

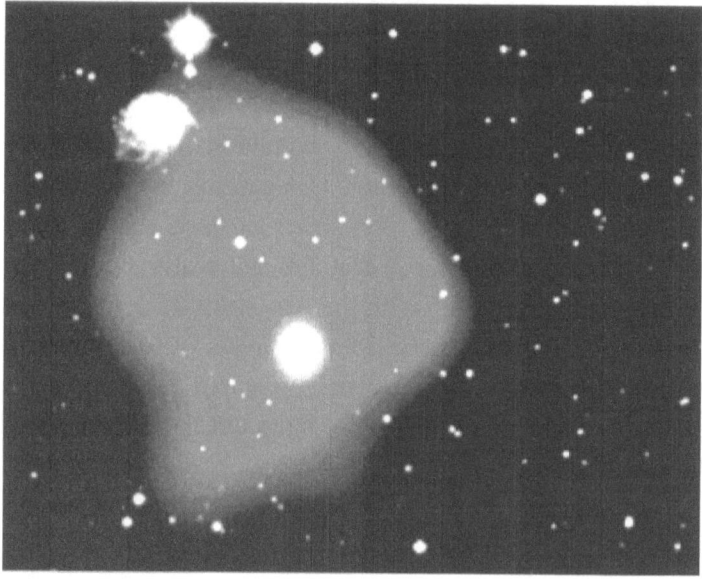

Figure 29: Hot X-ray Gas in a cluster of galaxies (http://imagine.gsfc.nasa.gov, 2010)

The cluster includes the galaxies and any material which is in the space between the galaxies. The force, or glue, that holds the cluster together is gravity: the mutual attraction of everything in the universe for everything else. The space between galaxies in clusters is filled with a hot gas. In fact, the gas is so hot (tens of millions of degrees) that it shines in the X-ray instead of visible light spectrum. In Figure 29, the hot X-ray gas (shown as the brighter cloud) lying between the galaxies is superimposed on an optical picture of the cluster of galaxies. By studying the distribution and temperature of the hot gas we can measure how much it is being squeezed by the force of gravity from all the material in the cluster. This allows scientists to determine how much total material (matter) there is in that part of space. (NASA, 2010)

Remarkably, it turns out there is five times more material in clusters of galaxies than we would expect from the galaxies and hot gas we can see. This is not just ordinary matter hidden from our view but rather something we cannot quite explain. Most of the "stuff" in clusters of galaxies is invisible and, since these are the largest structures in the universe held together by gravity, scientists have concluded that most of the matter in the entire universe is invisible. This invisible matter has been given the name, "dark matter."

There is currently a significant amount of ongoing research by scientists attempting to explain and understand exactly what dark matter is composed of, how much dark matter there is, and what effect it may have on the past and on the future of the universe as a whole. A complete understanding of dark matter, which may not even be possible, may provide some important clues to the origins and the history of the universe.

The discovery in 1998 that the universe was actually speeding up its expansion was a total shock to nearly all astronomers. To most people, it seemed so counter-intuitive, so against common sense. But the evidence that the expansion of the universe is speeding up is quite convincing. If you try to apply some common sense to this idea, it seems fairly logical.

If it is true that there is only a certain amount of "glue", or gravity, holding the universe together then as the objects become farther apart that rate of expansion should increase. I would further add that this rate of expansion should not increase linearly but rather exponentially since it is expanding in 3 dimensions.

Most of the evidence for the increasing expansion rate came from studying the distant Type Ia supernovae. This type of supernova is extremely rare and is a result of having a white dwarf star in a binary star system (see Figure 29).

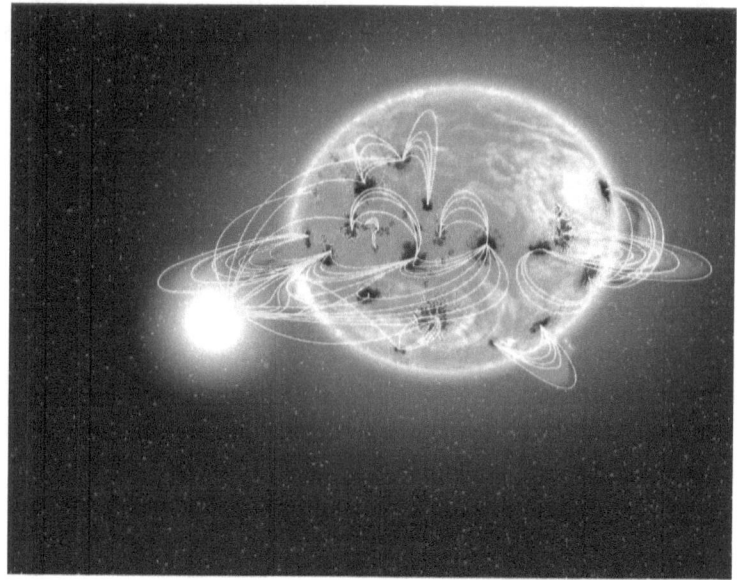

Figure 300: Highly Energetic Binary Stars called Polars (P. Marenfeld and NOAO/AURA/NSF)

In this type of binary star system, matter is believed to transfer from the relatively normal sized star to the white dwarf until the white dwarf attains a critical mass. This critical mass is known as the Chandrasekhar limit.

The Chandrasekhar limit is the point that limits the mass of bodies made from electron-degenerate matter, a dense form of matter which consists of nuclei immersed in a gas of electrons. The limit is the maximum non-rotating mass which can be supported against gravitational collapse by electron degeneracy pressure. It's named after the Indian astrophysicist, Subrahmanyan Chandrasekhar. Since white dwarfs are composed of electron-degenerate matter, no non-rotating white dwarfs can be heavier than the Chandrasekhar limit.

After a white dwarf attains the Chandrasekhar limit it undergoes a thermonuclear explosion. Because all white dwarfs achieve approximately the same mass before exploding, they all achieve the same luminosity and can be used by astronomers as "standard candles." Thus by observing their apparent brightness, astronomers can determine their distance using the $1/r^2$ law. This method of determining distance serves as a very clever cosmic yard-stick.

By knowing the distance to these supernovae, we know about how long ago they occurred. In addition, the light from the supernova has been red-shifted by the expansion of the universe. By measuring this red-shift from the spectrum of the supernova, astronomers can determine about how much the universe has expanded since the initial explosion. By studying many supernovae at different distances, astronomers can piece together a history of the expansion of the universe.

In the 1990's, two teams of astronomers, one at the Supernova Cosmology Project (at Lawrence Berkeley National Laboratory) and the other the High-Z Supernova Search (International team) were looking for distant Type Ia supernovae in order to measure the expansion rate of the universe with respect to time. They expected that the expansion would be slowing, which would be indicated by the supernovae being brighter than their red-shifts would indicate. Instead, they found that the supernovae were fainter than they estimated. Therefore, they concluded that the expansion of the universe was accelerating.

Dark Energy

We need to spend some time attempting to understand and explain dark energy. The likelihood of successfully explaining dark energy in this short section is about the same as me winning the Power Bowl without buying a ticket. But nonetheless, I'll attempt.

Not too long ago, by measuring the cosmic microwave background radiation, scientists concluded that the universe has a flat geometry on large scales. Because they could not observe enough matter in the universe, either ordinary matter or dark matter, to produce this flatness, the difference was attributed to what became known as "dark energy." This same dark energy is believed to have caused the acceleration of the expansion of the universe. In addition, the effect of dark energy seems to vary with time with the expansion of the universe. Basically, it appears to slow down and speed up at different times throughout its history.

There are some major differences between dark energy and dark matter. Astronomers believe that dark matter is present because of its gravitational effect on the matter that we see. There are also theories about the kinds of particles that dark matter must be made of. By contrast, dark energy remains a complete mystery to scientists. The name "dark energy" refers to the fact that some kind of stuff must fill the vast reaches of mostly empty space in the universe in order to be able to make space accelerate in its expansion. In this sense, it is believed to be a "field" just like an electric field or a magnetic field, both of which are produced by electromagnetic energy. But this analogy can only be taken so far, because we can readily observe electromagnetic energy via the particle that carries it, the photon. However, we have yet to be able to make this observation with dark energy.

The Cosmological Constant

Some astronomers have made comparisons of dark energy with

that of Einstein's famous *Cosmological Constant*. Einstein introduced this constant into his Theory of General Relativity when he saw that his theory was predicting an expanding universe, which was contrary to the evidence for a static universe that he and other physicists observed in the early 20th century. Additionally, an expanding universe went against his personal beliefs. The cosmological constant was able to balance the expansion dilemma and it made the universe once again static. Several years later, when Edwin Hubble's discovered the expansion of the universe, Einstein dismissed his constant. Einstein went on to call it the greatest blunder of his career. Later on, the cosmological constant became identified with what quantum theory calls the energy of the vacuum.

In the context of dark energy, the cosmological constant is believed to be a reservoir which stores energy. Its energy is believed to scale as the universe expands. Applied to the supernova data, the cosmological constant would distinguish the effects due to the matter in the universe from those effects due to the dark energy. Unfortunately, the amount of the stored energy required would be far more than what we observe and it would result in very rapid acceleration, so much that it would not be possible for the stars and galaxies to form.

Physicists have therefore suggested a new type of matter, "quintessence," which would fill the universe like a fluid which has a negative gravitational mass. This idea of quintessence reminds me somewhat of the old theories regarding ether. Ether, or luminiferous ether, was the hypothetical substance through which electromagnetic waves travel. It was originally proposed by the Greek philosopher Aristotle and subsequently used by several optical scientists as a way to allow propagation of light, which was believed to be impossible in empty space.

It was hypothesized that ether filled the whole universe and was a stationary frame of reference, which was rigid to electromagnetic waves but completely permeable to matter. Hooke endorsed the idea of the existence of the ether in his work *Micrographia (1665)*. Other philosophers of the 17th century, including Huygens, did the same. At the time of Maxwell's mathematical studies of electromagnetism, ether was still assumed to be the propagation medium and was imbued with physics properties such as permeability and permittivity.

In 1887, an experiment was performed by Michelson and Morley in an attempt to detect the existence of the ether. The experiment, named the Michelson-Morley experiment in honor of its authors, shocked the scientific community by yielding results which implied that ether did not exist. This result was later on used by Einstein to refute the existence of the ether and allowed him to develop Special Relativity without this artificial constraint. (Feynman, 1989) and (Krauss, 1995)

With flaws in the ether theories, it's not surprising that the new type of matter, quintessence, also has its issues. Recently discovered new constraints imposed on cosmological parameters by the Hubble Space Telescope data rule out the simple models of quintessence. Other dark matter possibilities being explored are topological defects, time-varying forms of dark energy, and a dark energy that does not scale uniformly with the expansion of the universe.

While all of these ideas are interesting, for me the most promising prospect is the unequal scaling of dark energy with the expansion of the universe. This is the theory that I'll focus most of my time and energy (possibly some dark) on. There are a couple reasons why I find this to be the most promising idea. One reason is that if space is analogous to the stretching of a rubber sheet, as suggested by Einstein in his General Relativity discussions and papers, then I think it is be reasonable to assume that space itself could be under different "tensions" at different points throughout the universe. Parts of this "rubber sheet" would be stretched and pulled more than other parts. This unequal stretching could reveal itself as the scaling of dark energy. Additionally, as the universe evolves and expands, different regions in space could also be affected differently by this unequal stretching of dark energy.

Data from Hubble provides supporting evidence that help astrophysicists to understand the nature of dark energy. This data will hopefully allow scientists to begin ruling out some competing explanations that predict that the strengthening of dark energy changes over time.

"Although dark energy accounts for more than 70 percent of the energy of the universe, we know very little about it, so each clue is precious," said Adam Riess, of the Space Telescope Science Institute and Johns Hopkins University in Baltimore. Riess led one of the first studies to reveal the presence of dark energy in 1998 and is the leader of the current Hubble study. "Our latest clue is that the stuff we call dark energy was relatively weak, but starting to make its presence felt nine billion years ago."

The Powerful Supernovae

To study the behavior of the dark energy a long time ago, Hubble had to peer far across the universe and back into time to detect supernovae. A supernova is the catastrophic death of a star, characterized by a massive output of energy. In the Milky Way, supernovae are relatively rare, with only a few notable incidences of historical supernova. In early times, many supernovae were probably topics of conversation and concern among the people who witnessed them. Around the universe, several hundred are observed and recorded each year, providing information about the formation of the universe and the objects within it.

Supernovae can be used by scientists to trace the universe's expansion. This has been described as being analogous to seeing fireflies on a summer night. Fireflies glow with about the same brightness (however there are some fireflies, like some humans, that don't have as much power going to their filaments). This light allows you to judge how the fireflies are distributed in the backyard by comparing their faintness or brightness, depending on their distance from you. The difference with space is that Hubble could not measure these ancient supernovae because they were too distant, and therefore they were too faint, to be studied, even by the largest ground-based telescopes.

Einstein first conceived the notion of a repulsive force in space in his attempt to balance the universe against the inward pull of its own gravity, which he thought would ultimately cause the universe to implode. At first, this seems to be a very reasonable and rationale thought. And it's certainly an idea that we should attempt to understand. This repulsive force that Einstein conceived could in fact have been the first idea similar to the notion of dark energy that we believe may provide the force that unequally scales and provides the kick that to the expansion and the increasing rate of expansion to the universe.

For several years, Einstein's cosmological constant remained a curious hypothesis and completely disregarded by most scientists until about 1998, when Riess and the members of the High-z Supernova Team and the Supernova Cosmology Project used ground-based telescopes and the Hubble telescope to detect the acceleration of the expansion of space from observations of distant supernovae. Astrophysicists came to the realization that Einstein may have been right after all. They concluded that there really was a repulsive form of gravity or some other repulsive force in space that soon after became dubbed *"Dark Energy."* What Einstein considered his greatest blunder, may not have been such a bad idea after all.

Over the past few years, astrophysicists have been trying to uncover two of dark energy's most fundamental properties:

1. What is the strength of dark energy?

2. What is its permanence (state of being permanent)?

Several new observations revealed that dark energy was present and obstructed on the gravitational pull of the matter in the universe even before it began to win this interstellar "tug of war."

Previous Hubble observations of the most distant supernovae known revealed that the early universe was dominated by matter whose gravity was slowing down the universe's expansion rate, like a ball rolling up a slight incline. The observations also confirmed that the expansion rate of the cosmos began speeding up about five to six billion years ago. That's when astronomers believe that dark energy's repulsive force overtook gravity's attractive grip.

The latest results pertaining to dark energy are based on an analysis of the 24 most distant supernovae known, most found within the last two years. By measuring the universe's relative size over time, astrophysicists have tracked the universe's growth spurts, much as a parent may witness the growth spurts of a child by tracking changes in their height on a doorframe. Distant supernovae provide the doorframe markings read by Hubble. "After we subtract the gravity from the known matter in the universe, we can see the dark energy pushing to get out," said Lou Strolger, astronomer and Hubble science team member at Western

Kentucky University in Bowling Green, Ky.

Further observations are presently underway with Hubble by Riess and his team which will hopefully continue to offer new clues to the nature of dark energy. (http://www.sciencedaily.com, 2010)

While nearly all of the research pertaining to dark energy happens in outer space, I believe some research should be conducted on the subatomic level. After all, if dark energy is so abundant in the universe then why should we not be able to find evidence of it here in our own backyard?

Perhaps some of the mysteries on the subatomic scale could be solved by carefully examining dark energy.

Chapter

Time Travel

*""Where there is no vision, the
people perish." Proverbs 29:18."*

I really wanted to get at least one really good Bible quote in this
book and thought, what better place than in the chapter
dedicated to time travel. The quote, however, has nothing to
do with time travel. Earlier in this book, I briefly discussed time
travel, and now it's time to go into a little more detail on the
subject. Due to the nature, this chapter will certainly be a bit
unusual; in fact, it may even be humorous to some readers. I'm
going to do my best to keep this chapter from going too far into the
science fiction arena, but at times it may.

For many years, time travel has been an area of great interest for all
types of people, including myself. For that reason alone, I thought
this subject was worthy of its own chapter. There are so many
questions out there about time travel and so many misconceptions.
Undoubtedly, the question that most commonly gets asked to
scientists and physicists is, *"Is time travel possible?"* Now make
sure you're sitting down, because the short and simple answer to
this question is, *yes.*

Maybe time travel is not possible using the ideas we're often
exposed to in science fiction movies, like flux capacitors or time
machines, but there are many good reasons to **not** rule out the
possibility of time travel completely.

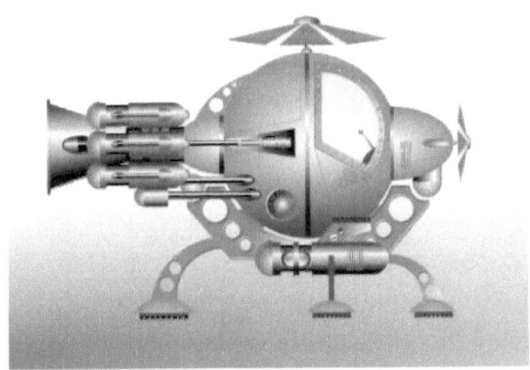

Figure 31: Fictional Time Machine (Loader, 2010)

The idea that time travel may be a possibility may surprise most of the readers, but it really shouldn't. After all, since you started reading this relatively short chapter, you've lost a few seconds that you may never get back, but probably wish that you could. While I don't expect a magical machine to be built like that pictured in Figure 30, time travel may become a reality by some other means.

Now, a better question to ask a scientist about the subject of time travel would be, *"Is time travel in both directions possible?"* I know that's a silly play on words from the original question but the answer to this question may surprise you as well.

Before I answer this newly rephrased question, I want to discuss some of the historical background on time travel. Newton once stated that the arrow of time only points in one direction. I don't remember the exact quote by Newton, however I wish I did, but it was something like that. I think that one simple sentence is a good place to start the debate on time travel.

Let's assume that Newton was correct. The Special Theory of Relativity showed us that while the arrow may only point in one direction, the size of the arrow may vary. Using that analogy, a big arrow would correlate to time passing fast. While a small arrow would equate to time passing slowly (I believe that the size of the arrow may not be as important as how the arrow is used). To be certain, Special Relativity clearly shows us that time travel is not quite as simple as originally stated by Newton.

Wormholes

Many very popular science fiction novels and movies have been made based on the subject of time travel. Some of these movies are my personal favorites. Believe it or not, actual scientists have even suggested that it may be possible to travel backwards in time via a theorized mechanism of outer space, and unintended consequence of General Relativity, called a wormhole. To understand this, there are a few questions to answer:

1. What is a wormhole?

2. Do they actually exist?

3. If they do exist can they be used for time travel?

I'll do my best to answer these and maybe even a few others questions. In physics and in science fiction, a wormhole is a hypothetical topological feature of space-time that would be, fundamentally, a "shortcut" through space and time. Although they are very popular in science fiction, there is no actual evidence that wormholes exist. For a simple visual explanation of a wormhole, consider space-time visualized as a two-dimensional (2-D) surface (see Figure 31). Now, if we "fold" this surface along a (non-existent) 3rd dimension, it allows us to picture a wormhole "bridge" (Please note that this image is merely a visualization displayed to convey an essentially unimaginable structure existing in 4 or more dimensions.). A wormhole is, in theory, much like a tunnel with two ends, each in separate points in space-time, created by space folded over onto itself.

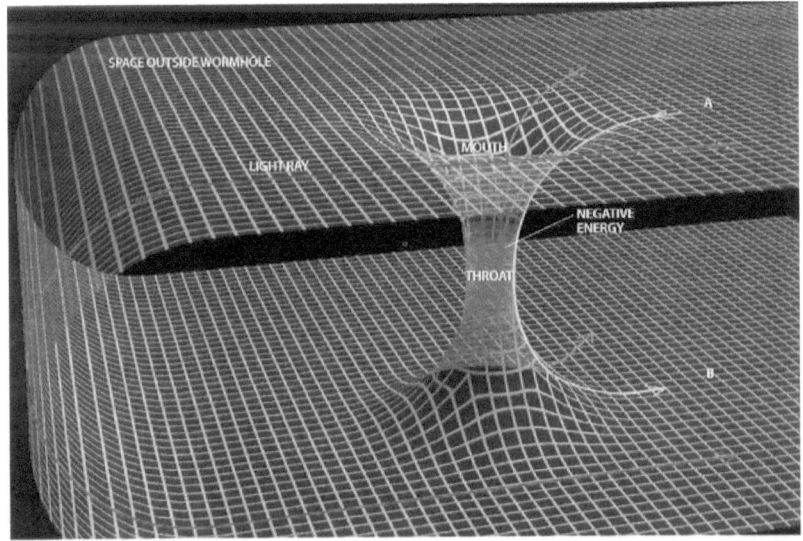

Figure 32: 3-D Wormhole (Wikipedia, 2010)

As previously stated, there is no observational evidence for wormholes, and, although wormholes are valid solutions in general relativity, this is only true if exotic matter can be used to stabilize them. Ok, so what is exotic matter? Exotic matter is a hypothetical concept of particle physics. It covers any material which violates one or more energy conditions or is not made of known baryonic particles. Such materials would possess qualities like negative mass (anti-matter) or being repelled rather than attracted by gravity.

The time flux analogy that I described in previous chapters for particles with mass can also be used to explain the idea of exotic matter. *You see, if a particle has a negative mass, it would in theory be exposed to a negative time flux*, as shown in Figure 32.

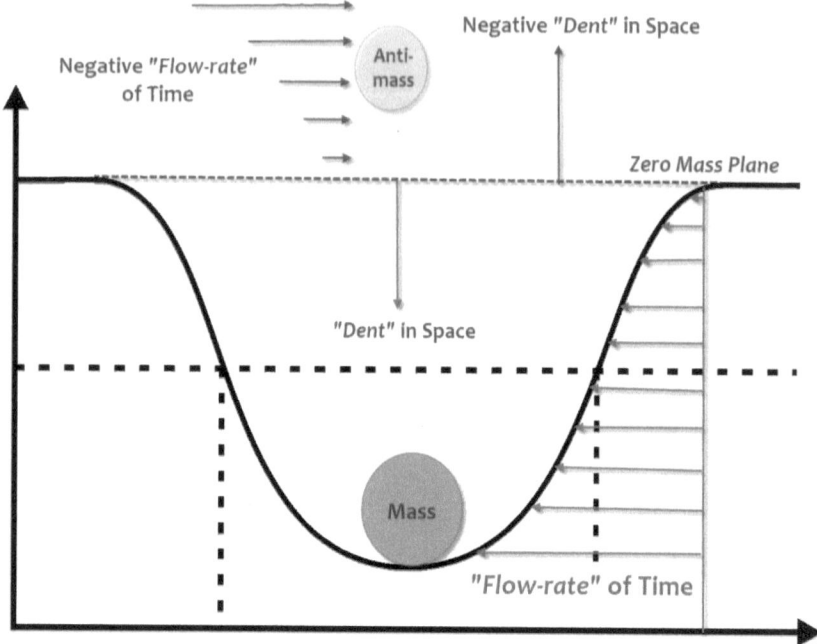

Figure 33: Anti-matter and Time

Figure 32 shows that objects with mass (the large ball at bottom of the well) create a dent in the fabric of space-time. Therefore, these objects are exposed to the flux of time. Anti-matter (called anti-mass in Figure 32) does not create a dent in space-time and lies above the zero mass plane. *Therefore this anti-matter is exposed to a negative flow-rate of time.* This figure is solely used as an illustration for this idea.

In practice, even if the wormhole could be stabilized, the slightest fluctuation in space would collapse it. If such exotic matter, that is, matter with negative mass does not exist, all wormhole containing solutions to Einstein's field equations are vacuum solutions, which require an impossible vacuum, free of all matter and energy. However, if anti-matter is found or created, a wormhole may become a reality.

It may be possible to create anti-matter by rapidly collapsing a region in space. The amount of energy required to do this would be extraordinary. *I believe it is quite possible the Large Hadron Collider is the testing ground for possibly creating anti-matter. I'm very skeptical in a 10 to 20 billion dollar science project being conducted solely for illumination. I think there's much more at stake at the LHC than may have been disclosed.*

By using current physical theories, one could conclude that wormholes might arise spontaneously, but would vanish nearly instantaneously, and would likely be undetectable. (Wikipedia, 2010) But with all the unknowns and uncertainties regarding wormholes, they tend to keep the sober physicist thinking that using them for time travel would not be possible.

Unlike most sober physicists, I certainly won't rule out the possibility that wormholes exists or could possibly be created. But for now I think it's best to consider how else time travel may be possible.

Dense Matter

There was a time, when I was very young and I thought that black holes as we know them didn't exist at all. Rather, I believed that what we thought of as a black hole was nothing more than a super massive and dense star or planet. This idea seemed to make sense to me. You see, one can calculate the speed necessary for an object to escape the earth's orbit. This term is called "the escape velocity." Simply put, this is the speed at which the kinetic energy plus the gravitational potential energy of an object is zero. It is commonly described as the speed needed to "break free" from a gravitational field. On the earth's surface, for example, if I threw a ball at about 25,000 miles/hr it could break free from the earth's orbit. It would also make me a very famous and rich baseball player.

Calculating various escape velocities is quite simple. The escape velocity for a planet or sun is directly proportional to the body's mass and inversely proportional to the body's radius. This means that a massive and small body (very dense) would have a large escape velocity. It's conceivable that a planet or star could be so dense that the escape velocity would be greater than the speed of light. What would an object like this look like to us? I suspect it would look similar to a black hole. We would see the effects of a large gravitational field revolving around an apparent emptiness in space. And as we approached closer to this type of black hole, we would simply burn up from the intense heat of the star.

To show the effect of super dense objects, we can consider stars with the same mass but with different sizes; you can see (intuitively) how the curvature of the equatorial plane changes. Shown in Figure 33, are pictures of physical and embedding diagrams of the equatorial plane. These embedding diagrams illustrate Schwarzshild's solution to the Einstein field equations (not discussed due to the complexity). Such a diagram can be extended to a black hole as well. (http://www.bun.kyoto-u.ac.jp, 2010)

STARS WITH THE SAME MASS, BUT DIFFERENT SIZES: HOW CURVED?

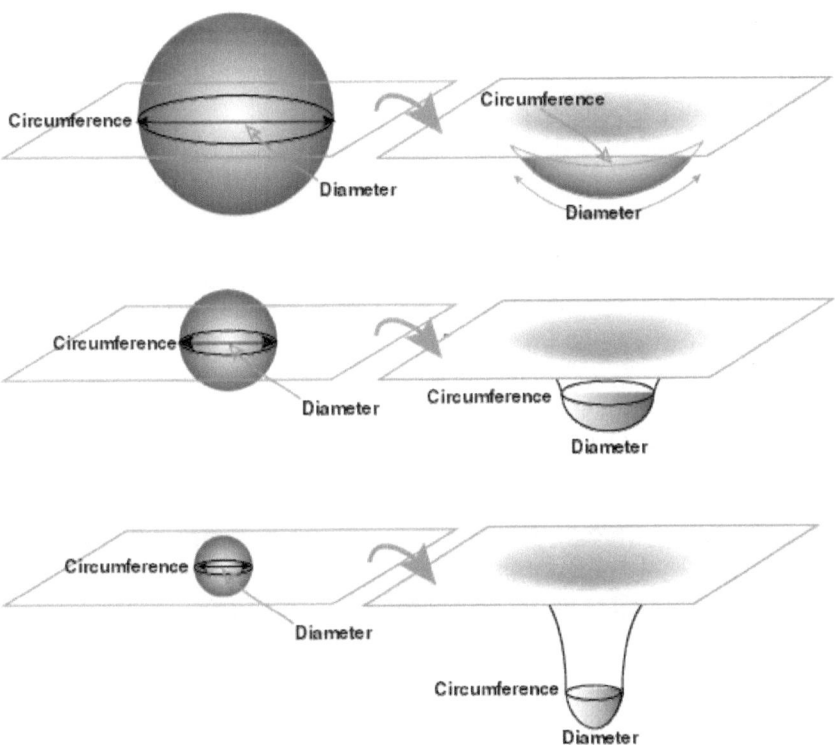

Figure 34: Stars with Same Mass but Different Sizes (Kips, 1994)

Why do I even mention this and what does this have to do with time travel? Well, it's a science fiction cliché to use black holes to travel though space and time. The story goes, if you dive into a black hole, you can pop out somewhere else in the universe, having traveled hundreds of light years in the blink of an eye, maybe even backwards in time.

But that's fiction. In reality, this probably won't work (assuming you could even survive being stretched in every direction with an incomprehensible force or burned into ashes by the super hot temperatures of a dense star). Black holes twist space and time, punching a hole in the fabric of the universe. There's a theory out there that if this happens it's possible a black hole can produce a wormhole. And if you enter a wormhole, you may pop out someplace far away, not needing to travel through the actual intervening distance.

As previously stated, while wormholes appear to be possible mathematically, in reality they would need to be made of theoretical forms of matter, mainly anti-matter, which may not occur in nature. The bottom line is that wormholes may not even exist. So when we finally perfect interstellar travel, we may not be able to take any shortcuts.

Evidence of Time Travel

I'll stop writing briefly about the possibility of reverse time travel and discuss what evidence, if any, we have to support it. Surely, if time travel were possible there would be some evidence we could examine today. After all, if at some point in history time travel becomes possible, why would there not be people from the future or relics from the future here with us here today? That seems crazy and it probably is, but I still want to consider it.

I once heard someone discuss that if a time machine were built, you would not be able to travel back in time to a point before when the time machine was built. That seems like a logical assumption but I'm not really sure that it's true or that there's any basis for that theory. For now, I want to focus on possible evidence of time travel and not this paradox.

Case 1: On Dec. 11, 2008, a team of Chinese tomb raiders announced they found a Swiss ring-watch, after removing the covering of the 400 year old Si Qing tomb, from Ming Dynasty, placed in Shangsi County. The watch appears similar to limited edition ring-watches made by LeCoultre and Rolex. Both were the first ring-watch models and started to be manufactured in the '50s. So far nobody could explain HOW that Swiss tiny watch got inside a 400 year old sealed tomb.

<u>Case 2</u>: A picture, originally posted in the Virtual Museum website, about an event that occurred in the 1940s, the reopening of the South Fork Bridge after flood in November 1940 in South Fork Bridge, Gold Bridge, B.C. Canada. A guy shown in the image, that seems to be very out place:

Figure 35: TIME TRAVEL: Strong evidence or major hoax???
http://www.abovetopsecret.com

His sunglasses, at first glance show side shields which could suggest a welding goggles model, but in the '40s, simply there weren't squared welding goggles. Also, his camera appears to have a huge objective lens. Before you say "so what?" be aware that in the 1940s there just weren't cameras with interchangeable lenses.

Case 3: Many people have been puzzled for several years by a scene in a Charlie Chaplin film which appears to show a woman talking on a mobile phone. The unusual thing is that the movie was made by Charlie Chaplin in 1928 – many years before mobile phones were invented.

As you watch the film, in the first 30 seconds there's a lady strolling by with her hand up to her ear which looks quite familiar in today's society. Most people who see it have the same reaction - it looks like she's talking on a cell phone.

Figure 36: "The Circus", 1928 silent film written and directed by Charlie Chaplin.

To really get the full effect of this, you should watch the clip for yourself on the internet.

Case 4: If people could travel back in time or see into the future what would they likely do with the information they would know in advance? For the most part, I consider seeing into the future to be in the same category as time travel because, in a sense, seeing into the future is the same as information travelling backwards in time (which may also violate Special Relativity). If we consider that information at some point has traveled backwards in time, I would like to think that people would use this information or knowledge for the good of humanity. For example, people with pre-knowledge of significant events could warn the world of upcoming disasters or prevent some other type of tragedy. But knowing a little about the nature of people, I think that most people would more likely use the information they gained for their own personal gain.

If for example, you knew the winning lottery numbers before the drawing, would you buy a ticket? I think most people would. After all, many people buy them with no pre-knowledge of the outcome. Some even consider a lottery ticket as an investment in their retirement plan. Statistically speaking, the odds in winning a lottery one time are about 1 in 100 million. This of course varies depending on how many numbers are used, how many numbers are picked, if there's a power-ball number, and a number of other factors.

There are so many different types of lotteries out there, but common to all of them is that the players all have a very low probability of winning. I recently heard that 14 people have won the lottery twice, a couple of these cases I was able to confirm. One woman reportedly won the lottery twice after dreaming of the winning numbers (http://www.snopes.com, 2010). The most remarkable story was of a Texas woman, Joan Ginther, who may just be the luckiest woman in the world.

According to a 2010 Huffington Post article titled, "Joan Ginther Wins Texas Lottery 4 Times", written by Paul J. Weber, this lucky woman has won the lottery four times. The chances of doing this are just one in 18 septillion. In case you (understandably) don't know how big 18 septillion is, it's 18 with 24 zeros at the end. And yet Ginther won $11 million in 1993, $2 million in 2006, $3 million in 2008 and $10 million in 2010.

The 63-year-old college professor is very elusive, even in her small hometown of 3,300 people. But the Times Market, where she bought her last two tickets, says business is booming. People often call from out of state and ask if the lucky store can ship tickets. They can't do that; that's illegal.

Although Ginther has never been investigated for her surprising number of wins, it seems to me that if there were a person who could rip off the lottery peeps, she'd be it. Ginther graduated from Stanford University and taught mathematics in California for 10 years, so she's smart and knows a bunch about numbers. Maybe she's figured out a foolproof lottery equation? Today, she lives in Las Vegas and possibly wipes her butt with dollar bills.

It seems reasonable that some people could win the lottery more than once, right? After all, there are so many lottery drawings every week and so many people play. But the odds of this feat are so small that it's hard to comprehend. These chances are so small that they defy any rational comparison. Someone would be about 100 billion times MORE likely to be struck by a meteorite than to win a standard state lottery twice. Forget about the case above where Ginther won the lottery four times. Something so wildly improbable should never have happened, as it was effectively impossible. And yet it did.

It turns out that there is a minor mistake in the above probability calculation and possibly even an error in the calculation by the Huffington Post; however I seriously doubt that due to the quality of their work. Nonetheless, it's a subtle, but very important error. Using simple probabilities for each of the steps (for example, a step would be winning a single lottery drawing) and multiplying them together would only be correct if the chances were completely random at each and every step. In other words, if someone had a 1 in 100 million chance of winning the lottery and does win, he still has a 1 in 100 million chance in winning the next time he plays. The one in 100 quadrillion odds only apply to the whole system. For example, a "Whole System" scenario would be something like: What are the odds of John Doe winning two consecutive lottery drawings?

This basic example shows one of the problems with statistics summed up in this humorous quote, *"People use statistics like a drunken man uses a light-post, for support rather than illumination."* Nonetheless, could defying these odds be evidence of information possibly flowing backwards in time? This question may seem crazy (and probably is), but we should still consider the possibility.

Higgs Singlet

We haven't really discussed the particle known as the Higgs Singlet, but this particle may hold the key to time travel. So what exactly is a Higgs Singlet? A Higgs Singlet is a theoretically proposed sort-of envelope which, if the numbers hold steady, could turn up alongside a Higgs Boson and its theoretical behaviour suggests it might actually exhibit time travel by appearing before the Higgs Boson which created or energized it appears.

Using this idea, scientists believe they are one step closer to creating time travel. American physicists from Vanderbilt University believe they may be able to use the Large Hadron Collider, the world's biggest atom smasher buried underground near Geneva, to send a type of matter called the Higgs singlet into the past.

The Higgs singlet may be able to jump through space and time, travel through a hidden dimension, and then re-enter our dimension forward or backward in time

But they're unsure if the Higgs singlet actually exists and whether the machine can produce it, according to a report by Live Science.

The Higgs singlet is related to another hypothesised particle called the Higgs boson, dubbed "God's particle" because it is associated with giving other particles mass. If the Higgs boson is created, the Higgs singlet may also appear, scientists say.

The Higgs singlet may be able to jump through space and time, travel through a hidden dimension, and then re-enter our dimension forwards or backwards in time, physicists Professor Thomas Weiler and graduate fellow Chui Man Ho believe.

"One of the attractive things about this approach to time travel is that it avoids all the big paradoxes," Professor Weiler said in a statement on research website arxiv.org.

"Because time travel is limited to these special particles, it is not possible for a man to travel back in time and murder one of his parents before he himself is born, for example."

"However, if scientists could control the production of Higgs singlets, they might be able to send messages to the past or future."

The singlet is a highly technical term to describe the particle that doesn't interact with matter in the usual way.

University of Sydney Associate Professor of Physics Kevin Varvell said the study was highly speculative, something the researchers themselves admit.

"From my reading of the paper, these guys themselves aren't going crazy over the idea of time travel," Professor Varvell said.

"They explicitly say we're not talking about time travel for humans, they're talking about potentially one might be able to send information through the production of these particles."

"But they're also saying that's very, very highly speculative as well."

He said it's one of many ideas that proposes using the collider and it is serious scientific work.

"But, again, I think we need to find the Higgs boson or something like it, before we can entertain other new particles being produced in association with it."

The Large Hadron Collider, which cost billions of dollars to build, has attracted plenty of controversy.

Before it started working, some feared it would create black holes and its operation was delayed several times due to a string of technical problems, including a liquid helium leak in 2008.

Metamaterial Manipulating Light

(Simon Hooper, CNN) -- New materials with the ability to manipulate the speed of light could enable the creation of a "space-time cloak" capable of masking events or even creating an illusion of "Star Trek"-style transportation, according to scientists in London.

The cloak, while currently only existing in mathematical theory, takes advantage of the potential properties of "metamaterials" -- artificial materials designed and manipulated at a molecular level to interact with and control electromagnetic waves.

Scientists have previously demonstrated that one possible use of metamaterials could be to render objects invisible by bending light around them. But Professor Martin McCall of Imperial College London says he has now extended the concept of invisibility to a cloak also capable of hiding events both in time and space.

"In some senses our work is mathematically quite closely related to the idea of invisibility cloaking," McCall told CNN. "It's just that we're doing it in space and time instead of just in space. It's added a new dimension to cloaking, quite literally."

In a paper published in the Journal of Optics, McCall said metamaterials made it theoretically possible to manipulate light rays as they enter a material so that some parts speed up and others slow down. This could create "blind spots" in time, masking an event. While the accelerated light arrives at a space before an event has happened, the rest of the light doesn't reach it until after the event.

"If you had someone moving along the corridor, it would appear to a distant observer as if they had relocated instantaneously, creating the illusion of a Star Trek transporter," says McCall. "So, theoretically, this person might be able to do something and you wouldn't notice."

Alberto Favaro, who worked on the project, compared the process to moving a pedestrian across a highway full of traffic by speeding up those cars already at or beyond the crossing point while slowing down the approaching vehicles.

"Meanwhile an observer down the road would only see a steady stream of traffic," said Favaro.

McCall said the theory could have practical implications in the future for quantum computing by opening up new possibilities for signal processing.

"If you have two channels that are carrying information, one of which has a continuous stream of bits on it, our technique can interrupt that stream and then process the other channel as a priority. So it can act as an 'interrupt without interrupt.' The original channels can then be seamed back together as if they'd never been interrupted."

The authors of the paper also joke that the "technology" would have its uses for criminals.

"A safe cracker would be able, for a brief time, to enter a scene, open the safe, remove its contents, close the door and exit the scene, whilst the record of a surveillance camera apparently showed that the safe door was closed all the time," they write.

The metamaterials necessary to create the perfect cloak are still many decades away, McCall said, while any prospect of upscaling the technology to conceal events even lasting a few minutes remains in the realm of science fiction because of the vast scale of the cloak that would be needed.

"Light travels at 100 million meters per second and in order to cloak it you'd need that many meters (of metamaterial), roughly speaking," he said.

But McCall said current optical-fiber technology could be used to construct a "poor man's cloak" capable of demonstrating "proof of concept" by imperfectly hiding events taking place over a few nanoseconds.

"We've provided a theoretical recipe and suggested how the experiment can be done. We believe the proof of principle experiment is available with current technology that experimentalist groups could achieve. It's up to the experimentalists to rise to the challenge," he said.

Ulf Leonhardt, a physicist working on the theory of invisibility technology at the University of St. Andrews in Scotland, said the paper was theoretically interesting but rejected the practicality of an experiment along the lines described by McCall.

"You'd need a very, very strong light source and it's not something you can make with commercially available devices that you can buy for a standard university laboratory," Leonhardt told CNN.

But he said research activities in the field of invisibility had exploded since he first published papers on the subject in 2006. While optical cloaking technologies remained a long way off, there have been some very promising experiments involving cloaking sound waves, he added.

"In acoustics you can definitely say this is working," said Leonhardt. "But it's still far away from being a practical technology for the optical range of the spectrum."

It's déjà-vu all over again –*Yogi Berra*

Have you ever been in a situation that seemed all too familiar to you, as if you've been there before? If so, you're not alone. This phenomenon is known as déjà-vu which comes from the French language, meaning "already seen." Basically, it's the experience of feeling sure that you've witnessed or experienced a new situation previously (an individual feels as though an event has already happened or has happened in the recent past). The reason that I personally can't ignore déjà-vu is because of the number of times it's happened to me and to virtually everyone who I know. Without going into great detail, there was one specific situation of déjà-vu in my life that certainly changed my life forever, as it potentially saved the life of one of my children.

Obviously, I can't just use my own experiences to conclude that déjà-vu is real and somehow information has traveled backwards in time. I'll need to back up my argument with some real facts. I was surprised to learn that up until about 1995, the United States Government was investing huge sums money and time into this very research. *Declassified documents related to the Stargate Project, revealed details of a $20 million research program sponsored by the U.S. Federal Government to determine any potential military application of psychic phenomena.* Although one Stargate viewer had been awarded in 1984 a legion of merit for determining "150 essential elements of information (...) unavailable from any other source" (Journal of Parapsychology, 1996) the program was eventually terminated in 1995, claiming a lack of documented evidence that the program had any value to the intelligence community.

The Government may still be doing some research today or perhaps they've given up, or perhaps they've successfully obtained the information they needed. To be certain, they were definitely looking for something. For this reason alone, I think this topic should not be brushed away without any regard. I think that there's a significant possibility that information can flow backwards in time, despite what our mathematical calculations tell us.

While it may be possible be that information can travel backwards in time, it would be a huge leap at this point to conclude that people or other physical objects could do the same. Pertaining to déjà-vu, I think we should all keep an open mind, but not so open that our brains spill out.

Little Green People?

Just to be certain that no stone is left unturned, I need to discuss (very briefly) the belief in extraterrestrial life if for no other reason than to completely discredit myself. I personally think that it would be arrogant to think that we are the only life-form out there in this incomprehensibly enormous universe. But do I believe that there's any _intelligent_ life out there? And do I believe that these aliens are visiting our planet? These are two completely different questions.

Let me first address the question about intelligent life, that's the easy one. It seems reasonable to conclude that there is a very special formula required to create intelligent life. Included in this formula are chemicals, conditions, and time. We've come pretty close to completing this recipe here on earth, but probably need a few million or billion years before intelligent life is actually achieved. But statistically speaking, the probability of intelligent life existing throughout our universe is high.

I'm reassured to know that I'm not alone in thinking that intelligent life exists outside our planet. A recently uncovered letter written by President John F. Kennedy to the head of the CIA showed that the president demanded to be shown highly confidential documents about UFOs 10 days before his assassination. The secret memo is one of two letters written by JFK asking for information about the paranormal on November 12, 1963, which have been released by the CIA recently.

Author William Lester said the CIA released the documents to him under the Freedom of Information Act after he made a request while researching his new book 'A Celebration of Freedom: JFK and the New Frontier.'

Alien researchers say the latest documents, released to Mr. Lester by the CIA, add weight to the suggestion that the president could have been shot to stop him discovering the truth about UFOs. For what it's worth, this is not a belief that I hold.

In the second memo, sent to the NASA administrator, the president expressed a desire for cooperation with the former Soviet Union on mutual outer space activities. The previously classified documents were released under the Freedom of Information Act to teacher William Lester as part of research for a new book about JFK.

Putting this presidential incident aside for a moment, is it possible that aliens have visited our planet? I personally don't think this is the case, but I still want to explore it. One could make a compelling argument that some of our technology is from the future or perhaps it's alien technology. We have, after all, made unprecedented leaps and bounds in the scientific world over the last hundred years.

It should also be mentioned that millions of people around the world have reportedly seen UFO's, some even claim to have seen aliens (many claim to have been abducted). Are all these people crazy? Not likely. The likelihood of all these people being wrong or being crazy is about the same as winning the lottery twice. But perhaps most of these people claim they saw a UFO or alien for attention. It's really hard to tell and I have not seen any evidence that convinces me beyond a shadow of a doubt of the existence of UFO's or aliens.

But let's consider this crazy idea for a moment: let's assume that at some point in the distant future, time travel becomes a reality. If this does happen, it would seem logical that these future people would want to travel back into time to understand the past. Perhaps much of our history will be lost during some great world war in the future. Could these sightings taking place today be nothing more than our own planet's inhabitants visiting us from the future?

It's an interesting idea to consider and at a minimum it may make a good topic for a future science fiction movie. Please just send me 15% of the profits.

End of Humanity

Chapter

10

"It is not, what a lawyer tells me I may do; but what humanity, reason, and justice, tells me I ought to do." -Burke, Edmund

J ust to make sure that I left no stones unturned (mainly so I never attempt to write another book), I decided to include a small chapter about the end of our race. Plus, I may as well discuss this since I've already discussed aliens and time travel. It may seem like a very dark topic (the end of humanity), but I want to make sure I don't end this book on a happy note. The book is after all mainly about your cat being dead. And if your cat dies, well maybe you should too. After all, is it really worth living after your cat dies? Prepare to be depressed.

I think there's little debate about whether the human race will end. To think that our species will survive for eternity is a bit arrogant and unrealistic. After all, from what we can tell, species have come and gone over the past millions of years on earth. The real questions to ask are:

1. How will humanity end?

2. And when will it happen?

For what it's worth, this is my two cents about the end of humanity, which is hopefully worth a tad more than just two.

Doomsday prophecies are a dime a dozen nowadays and have been popular for a few thousand years. There are many scenarios that could bring about the end of humanity, such as a giant meteor impact, a gamma ray blast from a nearby exploding star, a super volcano eruption, the end of the *"Keep your enemies close,* Mayan calendar, etc. I think the most *your friends closer and* likely cause; however, will come *bury your dead."* from within, self destruction from our own kind. *Personally, I'm far more concerned about the destructive nature of humans than the effects of a natural disaster impacting our environment.* While it's true that natural disasters pose a more lethal consequence, I think the likelihood of such event is far less than man's own destructive abilities. Since the likelihood of this is so much higher than a global natural event, I believe that man will eventually end the human race. But our destruction may not come as a result of another great world war or giant battle; it may come as a result of the greed of a few of the global elite.

The Wheel of Civilization

For a moment let's look at how a large system, business or a government works and compare it to some form of a simple machine. For the machine let's consider a simple cogwheel design, where there's one input shaft or gear, several cogs meshing together and one final output cogwheel that can produce a lot of torque or work, similar to the one depicted in Figure 36.

Figure 36: Cogwheel Machine

The input shaft gets its speed from some sort of fuel. For a business or government this fuel or power source is provided directly from the people or employees. The cogs are the real heavy hitters or big players in the business or government. Normal people like me and most of you are so insignificant that we're not even a sliver of wood on a single tooth of one of the cogs. We could be wiped from existence today and the machine would still have power, the cogs would turn and a product would be produced by the machine. It's real sobering to consider just how worthless most of us are to the big picture.

If an entire tooth were damaged the machine would still turn, and although it may operate a bit bumpy and it may not be not quite as efficient, a product would still be produced. If any entire cog were removed, the machine would stop for a period of time. But this machine (or business/government) is so developed that in virtually no time an adjustment would be made and the cog would be replaced and soon the same output would be achieved.

The output of these machines are money, power and property that go to a select and very secretive few that are so powerful they make the cogs seem insignificant. The key, however, to this whole process is a steady input power provided by the people of the society, company or government. If that initial power is not there, the machine stops and no output is produced. The machine is not literally fueled by people but rather by the money and services the people provide. This whole system is dependent on free or cheap labor and dependent on having money fed back into the system provided by the people.

This money comes in several forms; the most predominant forms are "hard taxes" and "soft taxes". Hard taxes are real taxes paid to the government for working, buying goods, buying property, buying services... etc. Soft taxes are not really taxes but rather a way designed to keep money out of your hands in order to be redirected back into the system. Often times these are passing fads that compel people to spend their hard earned money to keep up with the latest trends. Whether it's upgrading to the newest electronic gadgets or buying an outfit the hottest celebrity wears or endorses, the system goes through great trouble to keep your money flowing through the system. Eventually, it works its way out of the machine into the hands the select few. When money is no longer available from the people in the case of a real depression, money is created from thin air through borrowing, austerity measures and by cutting services. This borrowing of money keeps money going through the system while sucking fake money from the future and creating a larger and larger debt.

The only real way to affect the output of the machine is either to stop it from running or to completely disassemble the machine. To stop the machine from running is far more difficult than it sounds.

If the prime mover for the machine is money, time and work from the people than it would require a large percentage of the people to entirely stop feeding the machine. Even if this is done, the machine will find a way to keep moving. Maybe some sort of high levies or austerity measures would be put into place to force people to keep pumping money into the machine. The point is that the machine will not stop without a strong fight.

Disassembling the machine can be done by removing several cogs in the machine at the same time. The first step in this process is identifying who the cogs are. When the cogs are identified and removed from power the machine will try to quickly replace them. However, if enough cogs are removed the people may finally recognize how the machine operates and this awakening may be enough for the people to stop feeding the machine. This may prevent the machine from ever running or at least prevent it from running in the same fashion again.

The other way to disassemble the machine is to completely remove the foundation. This can only be done by finding the recipients of the output and stopping the flow of resources to these select few elites. This too is more difficult than it may seem. These elite individuals and families go through great lengths to ensure that they remain discreet. They're not the well known billionaires that you read about in money magazines or see on the news. These people are the real puppet masters and their wealth is in the trillions, not billions, and their power spans the entire globe. Removing these elites to shut down the machine is not a practical solution.

The perfect system may be some sort of cog/wheel machine where the cogs are people of great intelligence and integrity. And in this ideal machine the output would not benefit only a select few. In an ideal machine the people providing the power for the machine would also receive the benefits of the output. Many people think this is how a democratic/capitalist style of government functions: the harder you work, the richer your rewards. To some extent this is true. *But often times, it's a fixed game and in the end the dealer always wins.*

Some people will succeed in a democratic/capitalist system and these people will offer hope and encouragement to the millions who struggle to get by. The people who succeed are also part of the distraction that keeps the masses asleep. And this distraction and hope of success keeps powering the system. It's these everyday people who succeed that keep the masses from questioning the fairness and legality of the machine. After all, if these everyday people made it, then why can't I? All the while, the skimming off the top to benefit the world elites keeps millions of people down and forces them to depend of the system for their very survival.

At some point in the history of most countries, their system may have operated near perfection. But over time, sometimes slowly over decades or centuries, corruption crept in and altered the output. Corruption and greed can be blamed for nearly all the problems with civilizations and humanity. Until these core personality defects are corrected the cog/wheel machine will always morph into a corrupted system.

For Some People, Everything is Never Enough

Thinking back, there was a point in my life when I desired more than anything to become wealthy. And not just a little wealthy, but I wanted it all. I had a plan on how I was going to become a billionaire by the time I was 35. Needless to say, I didn't accomplish this goal. However, I'm certain I could have done so. Certainly, I could have come very close, but it would have been at the expense of many people in my life and it would have compromised my own morals.

Nowadays, I look at people with an inordinate amount of money much differently than most rational people do. You see, I firmly believe that every one of them either has a serious mental disorder, or they are completely ignorant, or they are the true face of evil. But they all fall into one of those three categories.

Let me explain how this works. Let's start by examining the people who are just plain evil. And believe me, there is true evil out there. This may seem harsh to suggest that some extremely wealthy

"If the American people ever allow private banks to control the issue of their currency, first by inflation, then by deflation, the banks and corporations that will grow up around them will deprive the people of all property until their children wake up homeless on the continent their Fathers conquered."

people are evil but let me explain. I've often wondered how anyone could go to sleep at night with millions or billions of dollars in the bank while so many people have nothing. To me, the very thought of this is completely incomprehensible. Money sitting in a bank does no good for humanity, in fact it hurts. It would be easier to understand and explain this if it was food these wealthy people were hoarding and not money. Imagine it was food for a moment. While people starve outside your castle, you sit on nearly all the food in the kingdom. Most people could clearly recognize that as evil, to watch people starve while you have so much.

Now let's talk about the ignorant people with wealth. Ironically, most of the people with extreme wealth think they're doing nothing wrong; after all they've "earned" their money, they're the job creators, and they donate some of their money and some of their time to charity. But if their fortune keeps going up or even stays the same, they have done nothing good, in fact they've only harmed society. But these wealthy men and women let their delusions persist and they think of themselves as saviors. I want to be very clear here: this chapter is not intended to and certainly will not change their ideas. Their beliefs are burned into their souls like the brandings on a cow. Hopefully this chapter will give ordinary citizens a new way to look at the people who some idolize and consider our social elite.

The simple fact is that the greed of a select few directly affects millions, perhaps billions of people around the world. None of the wealthy people will like this argument; it's sort of like arguing with a teenager. But they're all wrong. By sitting on so much money, it takes that much out of circulation for the rest of the people and indirectly causes many to remain poor. Some of the poor die as a result. Just because the wealthy are not stabbing anyone, doesn't mean their actions are harmless. For some time now the solution to this disparity has been to print more money. But in the current US model for every dollar that's created a dollar of debt is also born. This debt trickles down to those who are the most impoverished.

So is there anything wrong with a desire to have money? I don't think so. If fact, one could argue that much good could be done with it. Contrary to what some would have us believe, money is not the root of all evil, although insatiable greed may be. That's why the Bible says, "For the LOVE (desire or greed) of money is the root of all evil: which while some coveted after, they have erred from the faith, and pierced themselves through with many sorrows." (Timothy I 6:10)

Believe it or not, I'm a huge proponent of capitalism, not socialism. And I don't think that forced redistribution of wealth is a good idea (that's the basic concept of socialism). I also don't like the progressive tax system that we have in the United States. My ideas are based on the inherent goodness of people which I believe still exists to some extent in most people and they are based on true capitalism (not what we currently have in the United States).

You see, I believe that the people who are smart enough to accumulate so much money have the responsibility to put this money to good use, not to hoard it away from others. Most of us learn this simple lesson in kindergarten. Play nicely and share with others. But our consistent inability to do this is why I think we're one of the dumbest species on earth. We can let our own people die and feel no remorse whatsoever, while we eat our feasts like a bunch of fat cats. Even ants, with brains so small, appear to have more compassion for their own kind.

I did a back of the envelope calculation the other day and determined that the wealthiest *reported* 100 people in the world have nearly the same amount of money as the combined sum of every American (excluding the wealthiest 100). There are rumors that some families around the world are worth tens of trillions of dollars. I didn't include any of these rumors of wealth in my calculation.

The separation of wealth in this country is horrifying; in fact it's the highest of any industrialized nation.

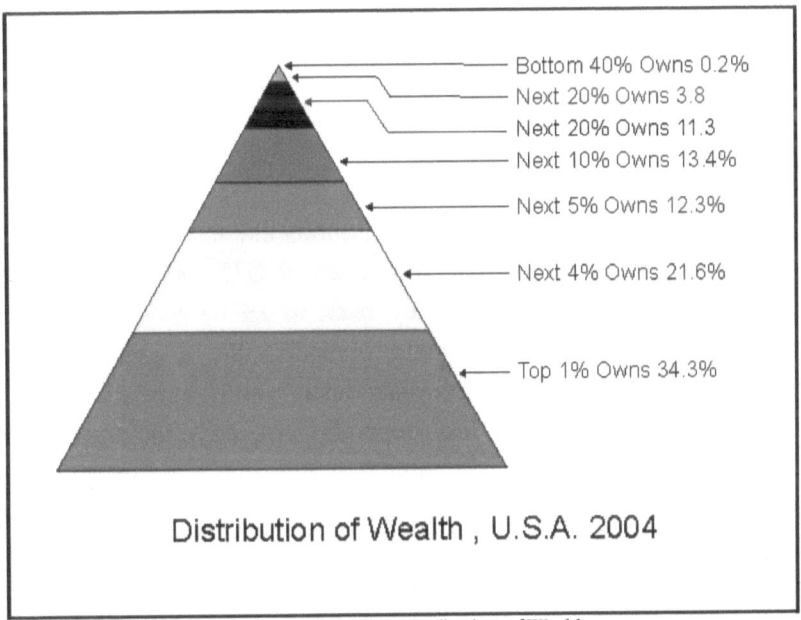

Figure 37: 2004 USA Distribution of Wealth

Figure 37 shows that 1% of Americans account for 34% of the wealth. This data is old and the numbers are far worse today. This separation between classes has been opening for decades. But yet we continue to buy into the whole system.

Next time you see a telethon, and think how nice it is that this famous celebrity is answering the phone asking for money from you, ask them the question, "Why the hell are you not contributing more? After all, you're the one with all the flipping money!" But I don't think celebrities are the real enemy we need to focus our energy on. They're just puppets that do what they're told, much like a small cog in my cog/wheel analogy. In a sense, they're nothing more than overpriced prostitutes.

I saw a show the other day about one of the world's wealthiest men. The show discussed how great it was that this man still lives like he did before he accumulated so much money. They described how he became wealthy by using simple investment techniques like *"Compounded interest is the most powerful force in the universe". – Albert Einstein.* investing in solid companies and holding certain stocks for a long duration of time. Sounds like a great plan, right? Well, imagine in 1960 that you invested $25,000 in the stock market (that was a lot of money back in 1960) and got a 17% return every year (most would consider this a good return, especially as of late). Fifty years later, you'd have about 100 million dollars (more or less depending on how the interest is compounded), assuming of course that you reinvested every dime you earned. So how then does one acquire 50 billion (or so) dollars and where does this money come from? I think the better question would be not where but from whom does this money come from? You already know the answer; it's simple: The money came from people like you and me.

The bottom line is that we need to stop being robbed by the wealthy people in our society. To do this, as a society we need to become much more conscience consumers. Every time we pull our wallet out, we should think about how each of the products we purchase was produced, who was affected by making them, how were the products marketed and where does the money that we're spending eventually end up. In most cases, the money winds up in the hands of a powerful few. It's time we get our hard earned share back.

Recently, we've seen people's life savings and retirement plans wiped out virtually overnight. Some were wiped out by criminals; some were wiped out legally by unethical people. But make no mistake about it, there's always a person or people behind the transfer of wealth. For some people, they'll never have enough money to satisfy their greed. For them, this is a disease that they're willing to kill or die for. For them, it's not about money; it's about control. And these people are obsessed with controlling every facet of our lives, from the food we eat to how long we should be permitted to live. How ironic is it that a carbon based life form would promote, propose, and attempt to enact a tax on carbon, the building block of our own existence? But as one of the social elite once said, "It's easy to lead sheep."

Most Americans, overall, are in debt to local, national and world banks, with their mortgages, loans, credit cards…etc. But the few people, with so much (all with no debt), are put on such high pedestals that people keep giving them more money and more power. Here's another dollar for your song, five dollars for your book, six dollars for your movie, and a hundred dollars for interest on the terrible loan you were kind enough to give me. If we don't come together and start helping our own species, we're doomed, and maybe rightly so.

Saving Humanity

So how do we start and what's the solution? Most people won't like the solution because it's hard, requires discipline, and requires faith. The faith part is easy; by simply arguing Newton's 2nd Law (equal and opposite), we know that God exists because of the sheer magnitude of evil in our world. But there's so much literature and information available to try to refute *His* existence. There are even new religions out there that are centered on oneself. These are all distractions to keep you occupied while you're being robbed, raped and pillaged.

Our society has become great at doing whatever it takes to make people feel good about themselves regardless of any direct or indirect effects. The bottom line is that everything has an effect (equal and opposite again). We all need to wake up and see what's really going on.

One of the problems is that there are so many distractions in our modern day world. We spend so much time watching meaningless television and focusing on worthless news. While our taxes and expenses go up and salaries go down, we're wrapped up in garbage television shows that keep us all talking around the water cooler. This is not by accident; rather it's part of a long term plan. The less we know the easier we can be controlled and exploited.

I was watching my kids enjoy a ride at an amusement park the other day and noticed some disturbing patterns. The ride they were on was the giant spinning swings, common to many parks. My kids went on the ride a few times, so I got to observe several rides (I enjoy roller coasters but generally say no to the spinney rides). I noticed that on each ride at least two people spent the entire time texting on their phones. They were so controlled that they couldn't even stop for three minutes to enjoy a park ride. It was a true shame watching these mindless robots, but again this was no accidental consequence of this technology. We live in a "got to have everything now" society. But if you keep your head in the sand you'll never see the truth.

The good news is that there is a solution to this problem. It's not easy, but it's obvious. And we all need to play a part in the solution. First, we need to look out for each other. Take a few minutes each day and do a good deed for someone else. This includes being charitable with your

"You let one ant stand up to us, and they all might stand up! Those "puny little ants" outnumber us a hundred to one. And if they ever figure that out... THERE GOES OUR WAY OF LIFE! It's not about food. It's about keeping those ants in line."

- Hopper, A Bug's Life

time and money. Next, put down your gadgets and wake up. The first thing you do when you wake up is open your eyes. Lastly, stand up and stop being robbed. Remember it's not about food, it's about keeping those ants in line (see quote above). Don't be afraid to step out of line from time to time.

If we sit by and let our world get taken over by a few elitist, we'll all be killed. It's up to us to take care of each other. *The sick, mental sociopaths controlling our world could care less about you and me. If we permit them to, they will end humanity.*

Summary

"I cannot conceive curved lines of force without the conditions of a physical existence in that intermediate space." Faraday

L ike the conclusion of a good mystery novel, this is where all the information gets wrapped together and begins to make sense. At least I hope that it makes sense. I'm certain that some of my logic seemed haphazard at times and my thoughts may have seemed out of place. There was, however, some thought behind the logic. Let me explain the flow of information in this book (you could call it the method to my madness) and the reasons why I chose the paths that I did. Yogi Berra once said, "When you come to a fork in the road, take it." And I did.

I wish I could make just one concluding statement that made this entire book come together and make sense, but unfortunately I can't do that. Here's the best statement I can make, *"Everything is, always has been, and always shall be energy."* It was over 100 years ago when Einstein showed that mass was energy. This bold new idea was a great start to what will someday be a complete unified theory.

I started this book searching to understand mass and that led me down many very different paths. Here's what we went over with some of the highlights and some of the new information discussed in each chapter:

Chapter 1: The first chapter was mostly front matter with introductions, goals and the basics of quantum mechanics. There were also discussions on how we perceive and interpret data. Sometimes in science, like life, we hold dear to ideas and principles for no particular reason. Most of the time when we ask, "why?", there's a very good reason. But sometimes, we enslave ourselves to these ideas. Perhaps if we step back and approach a problem from a completely different angle we may see what lies right in front of us: *those important things hiding in plain view.*

We also made an important assumption in Chapter 1: *There must be a point where we cannot divide particles any further. In other words, I assume that there's a smallest unit of matter or mass.* This idea is explored and amplified on throughout the book.

Chapter 2: This was a fairly short chapter where we went into the history and basics of quantum mechanics and chaos. After briefly describing these topics, one important question arose that steered a large portion of the rest of the book: *Could it be possible that the probabilistic nature of events on the subatomic scale may simply be the perception we have due to our own scientific limitations and the lack of understanding about the universe around us?*

Chapter 3: This painful chapter was a long review of many of the important rules and laws in science that we know today. We discussed Newton's Laws, conservation laws, Faraday's Laws, Maxwell's equations, fundamental forces, the Standard Model, the Higgs particle and a bit more. Mostly this was just an overview.

Chapter 4: This chapter is nearly all about Special and General Relativity and it includes topics like the "Twin Paradox", the "Equivalence Principle" and "String Theory". We also go into the hunt for a unified theory. Despite all the failed attempts in the past, I'm convinced that the search for a unified theory is not a total waste of time. Other important information and ideas are presented such as:

- No information is lost between observers in relative motion with each other.

- There is no concept of absolute rest.

In this chapter, we discuss how Einstein combined (postulated) the Equivalence Principle with Special Relativity to predict that clocks run at different rates in a gravitational potential, and light rays bend in a gravitational field. I went on to compare the ideas and theories in this chapter to show that General Relativity corresponds and agrees nicely with Kepler's Laws of Planetary Motion.

Chapter 5: In this chapter, I surmised that time may have been created out of the original mass (or energy) from the Big Bang. This time acts like a flux that passes through all objects with mass. Since mass-less observables do not create a dent in space, they are not exposed to this time flux and therefore no time will appear to pass for them. It was the stretching of space that originally created the flux of time. Time and space, in this regard, are therefore interdependent properties.

I also discussed the possibility that quantum mechanics may simply be part of a complete solution that will hopefully someday lead to a better understanding to the subatomic world and to our universe. A key to understanding the universe will be complete and accurate descriptions of space, time, and mass.

During the expansion following the Big Bang, mass was conserved, which includes the mass equivalent of energy. Also during the expansion process the combination of space and time was conserved. This was a new idea presented in the chapter: *During expansion, space was conserved, which includes the Space*

Equivalent of Time.

Out of this rapid stretching of space, time was created, or at least began to exist, as a dynamic measurable quantity.

Chapter 6: In the *Nature of Mass* chapter we built on some of the new ideas from earlier chapters and presented a few more new ones. We discussed the possibility that mass may be a quantized property. We went on to show how ripples in space may be able to trap energy and thereby create mass, or at least a perception of mass. Could it be possible that when the right amount and correct type of energy comes close enough together, the space around the energy traps it and creates some type of subatomic particle which we perceive as mass? Our perception of mass may be nothing more than the interaction of light with the ripples in space.

We then went on to describe an electron as a bundle of energy confined by the space around it. We went on to show how this may be connected to other types of energies or particles. You see, the force mediating particle that we are searching for may actually be just the ripples in space that we "feel." In other words, the Higgs particle itself may be the very structure of space that gives particles shape and mass.

When finished the chapter describing how this may relate to $E=mc^2$. When scientists describe this process, they simply see it as the transformation of mass to energy. But it could be possible that this process is just the unwrapping of space. To be consistent, energy has not been created nor destroyed in this process. The energy that was contained in the space around it has simply been released.

Chapter 7: This was a fairly short chapter that introduced the idea of space paths, leveraging on Fermat's Principle of Least Time.

Here, I discuss the possibility that quantum mechanics may simply predict the behavior of the space that the subatomic particles travel through. It may also be possible that the geometry and structure of space and time are actually fundamental properties. Using these ideas, I proposed that when particles or light travel down a space-time path, the potential for that path changes for the future light or particles that are shot. This makes that path less desirable for a period of time and forces a new route for the next particle.

Additionally, the rippling of space causes the forbidden regions that scientists observe. The electrons (or energy) are able to tunnel through the ripples in space rather than traveling over them. Thus the electrons appear to just show up in the new orbit rather than transiting smoothly between them.

The appearance of particles is nothing more than energy confined by the surrounding space. In other words, the frequency of light must fit inside the spatial constraints of the space ripples. The light must be in phase with or above the frequency of the space contours. If the frequency of light is below the threshold it cannot be deposited in the space ripples. The energy of the light will therefore not be deposited and it will not be able to free an electron.

Chapter 8: This chapter focuses on Dark Matter and Dark Energy and explores some new ideas regarding these widely unknown properties. I also discuss how this energy relates to the history and expansion of the universe. If it is true that there is only a certain amount of "glue", or gravity, holding the universe together then as the objects become farther apart that rate of expansion should increase. I further surmised that this rate of expansion should not increase linearly but rather exponentially.

There's still much more to learn in these areas. Perhaps some of the mysteries on the subatomic scale could be solved by carefully examining Dark Matter and Dark energy.

Chapter 9: Things start getting a little weird when we get to Chapter 9. Here I discuss time travel, UFO's and a few other controversial topics. I start by reviewing the time flux analogy that I described in previous chapters and use it to explain the idea of exotic matter. You see, if a particle has a negative mass, it would, in theory, be exposed to a negative time flux and therefore may be able to travel backwards in time. Anti-matter (called anti-mass in Figure 32) does not create a dent in space-time and lies above the zero mass plane. Therefore, this anti-matter is exposed to a negative flow-rate of time.

I believe it is quite possible the Large Hadron Collider is the testing ground for possibly creating anti-matter. I'm very skeptical in a 10 to 20 billion dollar science project being conducted solely for illumination. I think there's much more at stake at the LHC than may have been disclosed.

I end the chapter discussing various research programs established to determine applications and feasibilities of psychic phenomena and programs investigating UFOs.

Chapter 10: The final "main" chapter in the book focuses on the end of humanity, mainly added to end the book on a bright note. Personally, I'm far more concerned about the destructive nature of humans than the effects of a natural disaster impacting our environment.

The final chapter, Chapter 11, was not included in the summary for various reasons. When you understand the code in this chapter, you'll understand why it was excluded.

Throughout the book, I tried to present some basic background information on physics and what we currently know, or at least think we know, in science. Some of the sections like "time travel", "dark matter" and "string theory" were added to simply show the gamut of directions that science is taking and to discuss some of the new and popular topics in science. The bottom line is that there's still much that we don't know about the universe around us.

The meat of the book, the place where most of the new information is presented, is in the space, time, and mass chapters. The final secrets of the universe will be presented in the next chapter.

I believe that the key to a unified theory is to describe everything in the universe in terms of energy. Past attempts at unifying the four fundamental forces of nature has been hindered by many road blocks. Before we can describe and unify these forces, we must describe and attempt to understand energy at the most fundamental level.

Basically, I see mass as energy, confined by the geometry of space. Time is a property observed and felt by objects that dent space (objects with mass). Time interacts as a flux passing through objects with mass at a rate equal to the speed of light relative to the observer. Because of this relationship, time and space are intrinsically connected and cannot be separated. The three fundamental aspects of the universe that we need to understand are space, time and mass.

Space and mass are easy to visualize but not so easy to understand. I believe that time is the fabric upon which the framework of our universe, space, is laid. The speed of light, c is a perceived constant applied in some fashion to all mass-less observables, including time.

It's been shown that certain energies are quantized, that is, they have been shown to come in discrete quantized bundles. If this is actually true, then mass too should be quantized at its most fundamental level. After all, mass is simply energy. Using this logic, we could conclude that there must be a smallest unit of mass beyond which particles cannot be divided. This was one of the first assumptions that I made in the book and it may turn out to be correct.

If mass is actually a result from space, then space too must be quantized. If space is quantized, then so too must be time, since time is the fabric upon which space is laid. This is a disturbing thought and very hard for me to accept. Imagine that what appear to be smooth transitions of the second hand on a clock may actually be just summations of the smallest fundamental unit of time. This would paint a drastically different picture of the universe than the one we currently see.

If time comes in and passes by in discrete bundles then we know a little more about the Big Bang. Mainly, we know that time has not passed uniformly throughout time (if that makes any sense). If this were not the case, then we could not trace history back to a beginning. Or it may suggest that if we traced time back to a beginning it would approach a finite point but never actually arrive at it; hence there would have been no Big Bang. Since the passing of time is related to the speed of light, this suggests that the speed of light, c also has not always been constant and furthermore is not constant today. The changing of this factor is related to the stretching of space and may be too small to detect. But nonetheless, this and other factors we call constants, such as the gravitational constant, may not be constants after all.

The notion of bidirectional time travel cannot be dismissed, mainly regarding the passing of information. An object with a negative mass, or anti-matter, may be subjected to a negative time flux and therefore it may be possible for it to travel

"Man ... can go up against gravitation in a balloon, and why should he not hope that ultimately he may be able to stop or accelerate his drift along the Time-Dimension, or even turn about and travel the other way."

H.G. WELLS, The Time Machine

backwards in time. For humans beings, these objects or bits of information will appear to arrive and disappear from existence in an extremely short amount of time. In fact, they will appear to us for only one quantum of time before they travel into our past.

A quantum of time is a conclusion that I think we can reasonably draw. Most people who study quantization first learn of the quantized orbit, or allowed locations, of electrons in an atom. This theory explains why electron don't come crashing into the positively charged nucleus in a fraction of a second.

We showed that there were allowed locations that electrons could occupy. If we consider that the outer electron defines the radius of an atom, then this would imply that only certain sized atoms are allowed. Granted, this is a very small factor. When we consider that the average electron is separated from the nucleus by about 100,000 times the size of the nucleus we see just how small this factor is. But nonetheless, we can state that only certain size atoms are allowed.

Since we are restricted in the size of allowed atoms, we can conclude that only certain dents in space are allowed and therefore only certain masses or energies. The idea of quantized energies of electrons has long been known, but this type of quantization is different as it restricts space and time to only discrete intervals.

What we perceive as smooth transitions through space and time is nothing more than quantized summations of the smallest units of space and time. It's much like a reel-to-reel film. Each individual picture frame adds up when the movie is played to make a complete film.

The short chapter that follows, the last official chapter, is very brief but it contains a wealth of information. For fun, there's a code that may or may not contain any information of relevance. If the code does contain useful information, it is intended to be solved and understood by only a few. The clues to understanding this final chapter are everywhere and all around us but are as dynamic as the curves and forces on a roller coaster. Some people will understand this bumpy ride. Others will get sick from the motion; just keep hanging on.

The Coded Truth

"The society Shakespeare knew was heading for tremendous change, and he seems to have recognized that and written about it in a coded way. I understand those codes, I think. Redgrave"

Before I go too far in this section, I have to state that the code at the end of this chapter is intended for amusement or enlightenment purposes only. It may or may not reveal a message or clue. In fact, it may be nothing more than randomly generated characters. After all, what's the point of having a coded section in this book? What information could there possibly be to keep hidden from the masses and for what purpose would this secrecy serve?

"The First Clarke Law states, 'If an elderly but distinguished scientist says that something is possible he is almost certainly right, but if he says that it is impossible he is very probably wrong."

- Arthur C. Clarke

The answer to these questions could be quite simple. After all, there may not even be a code to decipher here. I believe, however, that there is certain information that some people are not ready to understand, comprehend and accept (information that may or may not be in the code). And if the general masses are privy to this information, the balance of the world and universe could be changed forever.

Consider for a moment that the first atomic bomb was constructed in 200 BC instead of during WWII. How different would changing this one event make our world today? Or what if the atomic bomb was never constructed? What if something we take for granted in every day, like for example the basic refrigeration cycle, was never discovered? Clearly these few examples show how drastically certain events, technology and knowledge have changed and shaped our history.

I want to go back and repeat a famous quote I mentioned earlier by Dr. Richard Feynman:

"If, in some cataclysm, all scientific knowledge were to be destroyed, and only one sentence passed on to the next generation of creatures, what statement would contain the most information in the fewest words? I believe it is the atomic hypothesis (or atomic fact, or whatever you wish to call it) that all things are made of atoms - little particles that move around in perpetual motion, attracting each other when they are a little distance apart, but repelling upon being squeezed into one another. In that one sentence you will see an enormous amount of information about the world, if just a little imagination and thinking are applied."

The simple fact that this quote was repeated twice in this book should show you how significant I believe it is. It's certainly worth reading more than once, perhaps even worthy of getting it tattooed to your lower back, as some sort of freakish, geekish tramp-stamp. Now, think about the simple title of this book: *Your Cat is Dead*. These are four words that to most seem meaningless and unimportant, unless you actually like your cat. But to a few other people, these words also contain a wealth of information.

If it was true that I had one grand new thought or truth about the universe that was so important that it could change history, I would want to somehow keep *some* people from the information. It's important to remember that not all people are innocent and have benevolent thoughts and intentions regarding mankind. Some people are downright evil. Therefore, if I had some spectacular insight to the universe, I would want it protected by some sort of encrypted code.

I'm pretty certain, this idea would not be quite as condensed as Dr. Feynman's one sentence but this thought would also contain an *infinite* wealth of information. I hope someday, someone solves all the grand problems of the universe. For now, I will leave you with a code that may answer every question or it may prove to be meaningless. I believe it will be solved by a future generation, long after I leave this unforgiving world.

I attempted to design a code so that computers wouldn't be able to crack it; either that or it's just gibberish. As I mentioned before, it may be nothing or it may answer everything. But regardless, it will take some intelligent being, a being with a pulse to do so. It's probably not a terribly complex code, that's not my style. But when cracked, then you (and you alone) are capable of understanding the information. I doubt that any actor, politician, banker or Wall Street executive will ever solve this puzzle on their own. They simply do not have the right tools in their tool-bag and their bag is two and a half sizes too small, just like the heart of the Grinch.

"One of the saddest lessons of history is this: If we've been bamboozled long enough, we tend to reject any evidence of the bamboozle. The bamboozle has captured us. Once you give a charlatan power over you, you almost never get it back."
- Carl Sagan

Don't look at this as an intellectual challenge that once completed you share the information with your colleagues. And don't look at it as a secret that once revealed you sell to others. But rather it's the key to opening the door into an exclusive club that you've earned the privilege to join. Don't give away your precious key because then you can never return.

What's revealed behind this code could be the key to our world and universe. All the hints and answers to help solve the code are in this chapter and in the rest of the book. Before you attempt to crack the code, I want to leave you with one final quote that will hopefully make sense to you and help with the challenge at hand:

"The important thing in science is not so much to obtain new facts as to discover new ways of thinking about them." - Sir William Bragg

The Code

D6EIT7EKCD*R8M*FK66RKI9FO2P<u>KD</u>5ALSHG9BSKU
NRK*266*EIM8PKD9MKGKULEQJKQ9LS

KFU7DMH6MPSHQ266SC2CH
O9GYHBJLS97QIAVC2EOF26
6CUDIAZ3DXLEKT3HHC7*FV*4UDH4CSQIYK
HPA3GWPHP<u>V</u>5BJG7VSOMZ266EZRHT7I
XRX7YF7<u>OWN</u>QMFZ6

50

Glossary

Most glossaries simply contain terms that are used but usually not described in their book. This one is a little bit different. While it does contain many uncommonly used terms, it also contains many terms and phrases that I added for further educational purposes. Even if you do not need to look up a term, I recommend at least browsing this section.

Accelerator: Accelerators are ring-shaped or linear devices that accelerate charged particles. More powerful than any other microscope, high-energy accelerators allow physicists to study matter at the smallest scale human beings have ever seen, exposing the quarks inside a proton. At the same time, high-energy accelerators are attempting to produce collisions that recreate the conditions of the early universe.

AGS (Alternating Gradient Synchrotron): An accelerator based at Brookhaven National Laboratory (BNL).

AMANDA (Antartic Muon and Neutrino Detector Array): A Cherenkov detector, embedding in the ice located at the South Pole, designed to look at very high energy neutrinos.

Antimatter: All particles of ordinary matter (electrons, protons, neutrons) are believed to have anti-matter partners that appear identical in all respects (e.g. mass, spin) except that they have the opposite electric charge. It's believed that in the Big Bang equal quantities of matter and antimatter were created.

Antiquark: The anti-matter partner of the quark.

Antiparticle: The generic term for an anti-matter partner of a particle.

Antiproton: The anti-matter partner of the proton.

ATLAS (A Toroidal LHC Apparatus): This multi-purpose experiment is currently under construction for use as a detector at the Large Hadron Collider (LHC) in Geneva Switzerland.

B-Meson: A meson containing a bottom (b) quark, and one lighter anti-quark. The b quark is the second heaviest quark, and is found only at particle accelerators. Only the top quark is heavier.

Barn: A unit of cross-section, a barn is equal to 10^{-28} m^2.

Baryons: A hadron composed of three quarks. Examples include the protons and neutrons found in ordinary nuclei.

BEPC: Beijing Electron Positron Collider

Big Bang Theory: The Big Bang theory generally refers to the idea that the universe has expanded from a primordial hot and dense initial condition at some finite time in the past and continues to expand today.

Bilderberg Group: The Bilderberg Group, Bilderberg conference, or Bilderberg Club is an annual, unofficial, invitation-only conference of approximately 140 guests, most of whom are people of influence in the fields of politics, banking, business, the military and media. The conferences are closed to the public and the media, and no press releases are issued. Because of its secrecy and refusal to issue news releases, the Bilderberg group is frequently accused of political conspiracies. This outlook has been popular on both extremes of the ideological spectrum, even if they disagree on what the group wants to do. Some accuse the Bilderberg group of conspiring to impose capitalist domination, while other groups have accused the group of conspiring to impose a world government and planned economy.

BNL (Brookhaven National Laboratory): A national laboratory located in Long Island, New York. This multidisciplinary laboratory, operated by the Department of Energy, is the site of the AGS and RHIC (relativistic heavy-ion collider) accelerators.

BooNE (Booster Neutrino Experiment): A proposed experiment to be based at Fermilab. It is the expanded version of the Mini-BooNE experiment, which is currently under construction.

Branching Ratio/Fraction: When a particle decays, it often can decay in several ways. The likelihood of it decaying to a particular mode is known as its branching ratio for that decay mode.

BTeV: A proposed experiment to take place at the Fermilab Tevatron. It would study B-Meson decay in detail.

CDF (Collider Detector at Fermilab): This is one of the two large multi-purpose experiments located at the Fermilab Tevatron. Along with D0, it discovered the top quark in 1995.

Center-of-Mass Energy: In particle collisions, this is the energy that can go into making new particles. For a collider experiment where two beams of equal energy collide head-on, this is simply the sum of the energy of the two beams. In fixed target experiments, in which a beam of particles strikes a stationary target, the center-of-mass energy is significantly less than the sum of the energies of the two colliding particles.

CERN (European Organization for Nuclear Research): A particle physics laboratory in Geneva, Switzerland that is supported by 20 member states. It is the site of the Large Hadron Collider (LHC), which is currently under construction. Over the last fifteen years, the experiments at CERN have provided particle physicists crucial information about electroweak unification.

CESR (The Cornell Electron-positron Storage Ring): A machine, located at Cornell University, funded by the National Science Foundation, that collides a beam of electrons and a beam of positrons of equal energy. The total Energy is adjustable from 9^{-12} GeV.

Chandrasekhar Limit - In a binary star system, matter is believed to transfer from the relatively normal sized star to the white dwarf until the white dwarf attains a critical mass, known as the Chandrasekhar limit. The Chandrasekhar limit is the point that limits the mass of bodies made from electron-degenerate matter, a dense form of matter which consists of nuclei immersed in a gas of electrons. The limit is the maximum non-rotating mass which can be supported against gravitational collapse by electron degeneracy pressure. It's named after the Indian astrophysicist, Subrahmanyan Chandrasekhar.

Charmed Particle: Any particle that contains the exotic charmed (c) quark is known as a charmed particle.

Cherenkov Detector: Light travels more slowly in materials, such as water, than it does in a vacuum. No particle can travel faster than the speed of light in a vacuum. Particles that travel faster than the speed of light in a given material emit a cone of light, in the same way that objects exceeding the speed of sound emit a sonic boom. Detectors that use this light to detect subatomic particles are known as Cherenkov detectors.

CLIC (Compact Linear Collider): A proposed linear collider under study by an international group based at CERN. The proposal is for a collider that could reach energies of 5 TeV.

CKM (Charged Kaons at the Main Injector): A proposed fixed target experiment to take place at Fermilab. Its goal is to measure the decay of charged K-mesons to pions and neutrinos to study the parameters of the CKM mixing matrix.

CMS: (Compact Muon Solenoid): This large, multi-purpose detector is currently under construction for use at the Large Hadron Collider (LHC) in Geneva, Switzerland.

Color: A property of quarks and gluons. Gluons can exchange color between quarks and other gluons. This process is the origin of the strong force. It has nothing to do with the color perceived by the human eye.

Cosmic Rays: Any of the particles from outer space that are continuously colliding with the Earth's atmosphere. They are mostly protons, with some nuclei, electrons, and photons. Their interactions with the atmosphere produce a variety of particles, including pions, muons, and neutrinos.

Coupling Constant: The parameter that describes the strength of a given force.

CP Violation: CP is a symmetry that relates particles to anti-particles. CP violation occurs when there is a difference in the way that particles and anti-particles interact. It is believed to be necessary for the excess of matter over antimatter we observe in the universe. This phenomenon is currently being studied in detail at BaBar and BELLE.

Cross-Section: a measure of the likelihood of a given process occurring at an accelerator. The idea is that two objects with a larger cross-sectional area are more likely to hit one another. So, larger cross-sections mean that a process is more likely to occur. Cross-sections are measured in barns, 10^{-28} m^2. A barn is an extremely large cross-section in particle physics. Many interesting cross-sections are measured in pb (picobarns), which are equal to 10^{-12} barns.

Dark Energy: A poorly understood (and non-luminous) substance that exerts a pressure that tends to accelerate the expansion of the universe. This dark energy counteracts gravity's natural tendency to slow the expansion of the universe. It is particularly important to understand this substance, as it appears to make up the majority of our universe. Dark energy is also often referred to as the "cosmological constant". It is distinct from dark matter, which gravitates in the same way as ordinary matter, but is not luminous.

Dark Matter: Astronomical measurements indicate that luminous matter, such as our sun, makes up only a small percentage of the total matter in the universe. The missing mass that makes up the remainder is known as dark matter.

Decay: Exotic particles produced at accelerators are often very short-lived, and can transform into lighter, less exotic products, such as electrons and photons. This process of transformation is known as decay.

DESY: A laboratory located in Hamburg, Germany. It is the site of the HERA accelerator.

DOE (Department of Energy): One of the principal federal agencies supporting research in the physical sciences in the United States. Through its Office of Science it provides approximately 90% of the support for High Energy and Nuclear Physics.

D0 (D-Zero): One of the two large multi-purpose experiments located at the Tevatron at Fermilab. Along with CDF, it discovered the top quark in 1995.

Double Beta Decay: The process in which a neutron in a nucleus transforms into a proton by emitting an electron and an anti-neutrino is known as beta decay. When two neutrons undergo this transformation simultaneously, the process is called double beta-decay. Of particular interest is the hypothetical process of *neutrinoless double beta decay*, in which the neutrons transform by emitting two electrons, but no anti-neutrinos.

Electron: A fundamental constituent of matter. Along with protons and neutrons, electrons are the building blocks of atoms. They have negative electric charge.

Electroweak Symmetry Breaking: Although electromagnetism and the weak force have the same strength at high energies, electromagnetism is much stronger than the weak force in our everyday experience. The process by which a single unified electroweak force becomes two separate forces is electroweak symmetry breaking.

Electroweak Unification: This theory describes two of the four fundamental forces, elctromagnetism and the weak interaction (responsible for nuclear decays) as a single force at high energy.

eV (electron volt): A unit of energy equal to the amount kinetic energy an electron gains after being accelerated through an electric potential of 1 Volt. It can also be used as a unit of mass by applying Einstein's relation $E=mc^2$.

fb (femtobarn): 10^{-15} barns. A unit used to measure cross-section. The inverse of this unit is used to measure integrated luminosity.

Fermilab: A laboratory, operated by the Department of Energy, located in Batavia, Illinois. It is the site of the machine that currently operates at the highest energy in the world, the Tevatron.

Fermion: The term for a particle with half-integer spin. Examples include the quarks and leptons of the Standard Model.

Fixed Target Experiment: An experiment in which a single energetic beam strikes a stationary target such as a block of metal. This is to be contrasted with collider experiments, in which two energetic beams collide head on.

Flavor: Flavor has two different meanings. For leptons, it is the label used to differentiate the three generations. That is to say, electrons, muons, and taus are said to have different flavor. For quarks, it is a label that discriminates between all quarks of different masses. There are six known flavors of quarks: up(u), down(d), strange(s), charm(c), bottom(b), and top(b).

Galactic Halo: the region of dark matter that surrounds the visible portion of a galaxy.

Gamma Rays: photons of high energy. The most energetic forms of light are known as gamma rays.

Gauge Bosons: The particles that carry the forces of the Standard Model: electromagnetism (photons), the weak force (W and Z-bosons), and the strong force (gluons).

Gaugino: The generic term describing the hypothetical superpartner of any of the gauge bosons of the Standard Model.

General Relativity: Einstein generalized his theory of special relativity to include gravity, and called it a general theory of relativity. It showed that apples fall to the ground because the Earth's mass curves the adjacent space-time, forcing apples to move in a special way-towards the surface of the Earth. It has proved, however, extremely difficult to unify General Relativity with quantum mechanics. String theory is currently the best hope for a theory of quantum gravity.

GeV (Giga-electron Volts): 10^9 electron volts

GLAST (Gamma Ray Large Area Space Telescope): A satellite-based experiment that searches for gamma rays.

Gluons: The spin-one particle that carries the strong force.

Gluino: The hypothetical superpartner of the gluon.

Gran Sasso: An underground laboratory located near Rome, Italy. It is the site of dark matter detection experiments, a double beta decay experiment, and neutrino detection experiments.

Gravitational Lensing: According to Einstein's theory of general relativity, the presence of matter can warp space-time. This warping of space can affect the path which light-rays follow, much in the same way that a lens does. This is known as gravitational lensing.

Graviton: The as-yet unobserved spin-two particle that carries gravity.

HEPAP (High-Energy Physics Advisory Panel): The committee that advises the Department of Energy and National Science Foundation on the particle physics program.

HEPAP subpanel: HEPAP appoints a subpanel every few years to have intensive and detailed discussions on specific problems. This subpanel was appointed in March 2001 to discuss the future of the particle physics program in the US.

Hadron: Any particle made out of quarks and/or anti-quarks. Protons, neutrons, and anti-protons are examples of hadrons.

Hadron Collider: A machine that accelerates hadrons to high-energies in two beams, and then collides the beams head-on.

HERA: An accelerator, located at the DESY laboratory that collides 920 GeV protons and 30 GeV electrons.

HERA-B: An experiment located at the DESY laboratory. HERA-B used the 920 GeV proton beam of HERA (data acquisition until 2003).

HERMES: An experiment located at the DESY laboratory. HERMES uses the polarized electron beam of HERA.

Hidden Dimensions: Hypothetical additional dimensions of space-time, either a classical dimension in which particles can move, or a quantum dimension that converts a force particle into a matter particle and vice versa.

Higgs Boson (H): An as yet undiscovered particle that appears when sufficient energy is imparted to the Higgs Field.

Higgs Field: A hypothetical medium that permeates space and time, and is thought to impart masses to all of the particles of the Standard Model. Without the Higgs Field, all particles in the Standard Model would have no mass.

Higgs Singlet: A theoretically proposed sort-of envelope which, if the numbers hold steady, could turn up alongside a Higgs Boson and its theoretical behaviour suggests it might actually exhibit time travel by appearing before the Higgs Boson which created or energized it appears.

High-Energy Physics: Because particle physics often requires high-energy particles to probe short distances, particle physics and high-energy physics are often used interchangeably.

HiRes (The High Resolution Fly'sEye): A cosmic ray detector based in Utah. This detector is especially sensitive to the highest energy cosmic rays.

Homestake: A former gold-mine located in South Dakota that is the proposed site of a National Underground Sciences Laboratory. This mine also contained the experiment that first detected solar neutrinos.

Integrated Luminosity: This is a measure of the total data collected at an accelerator. It is the intensity of the accelerator, summed over some specified time period. A $barn^{-1}$ (inverse barn) of data will give one event for a process that has a cross-section of one barn. A $picobarn^{-1}$ (a much larger amount of data) will give one event for a much rarer process that has a cross-section of one picobarn.

JHF(Japan Hadron Facility): An accelerator currently under construction for nuclear physics experiments that will produce a very intense 50 GeV proton beam. It can also be made to generate an intense beam of neutrinos.

JINR: The Joint Institute for Nuclear Research in Dubna, Russia.

JLC (Japanese Linear Collider): A development effort for a next-generation linear collider based at KEK.

K2K: This experiment aims a neutrino beam from the KEK accelerator to the SuperKamiokande detector. It is searching for neutrino oscillations.

KamLAND: An experiment based at the Kamioka mine in Japan that is designed to look at anti-neutrinos produced by nearby nuclear power plants.

K-Meson: This is the name for a meson that contains the exotic strange (s) quark and either an up or down quark. These particles are also known as Kaons.

KEK: A high energy physics laboratory located in near Tokyo, Japan. It houses the BELLE experiment, as well as the accelerator used in the K2K experiment.

K0PI0: An experiment that searches for a rare decay of uncharged K-mesons into neutral pions. This decay provides valuable information about the CKM Mixing Matrix and CP violation.

LBNL (Lawrence Berkeley National Laboratory): A Department of Energy facility located in Berkeley, California, with a broad-based program of basic and applied research.

LEP (Large Electron Positron Collider): This collider, operated from 1989 until 2001 at the CERN laboratory in Geneva, Switzerland, was used to make detailed studies of the weak force. The 27 km tunnel that housed this accelerator will now house the LHC.

Lepton: Along with quarks, leptons make up all known matter. Unlike quarks, leptons do not participate in strong interactions. Neutrinos and electrons are two common examples. There are three flavors of leptons: electron, muon, and tau.

Lepton Flavor Violation: In the Standard Model, leptons do not change flavor. For example, a muon would never turn into an electron. If this were observed, it would be a signal for new physics. Evidence for Neutrino oscillations already indicate that lepton flavor violation may occur.

LHC (Large Hadron Collider): A 14 TeV proton-proton collider under construction at the CERN laboratory in Geneva, Switzerland.

Linear Collider: An accelerator that accelerates two beams of charged particles in a straight line, and then collides them head on.

LLNL (Lawrence Livermore National Laboratory): A Department of Energy facility located in Livermore, California, with a broad-based program of basic and applied research.

LSST (Large-aperture Synoptic Survey Telescope): A proposed telescope, whose possible applications range from the study of dark matter to searching for asteroids that could collide with the Earth.

Luminosity: This figure of merit for an accelerator quantifies the intensity of the beams. It is directly related to the number of events produced at the machine.

MECO (Muon to Electron Conversion Experiment): A proposed experiment to look for the conversion of muons to electrons in the presence of nuclei. This would be a signal of lepton flavor violation. meson: A hadron composed of a quark and an anti-quark.

MeV (Mega-electron Volts): 10^6 electron volts.

MiniBooNE (Mini- Booster Neutrino Experiment): This experiment, based at Fermilab, represents the first stage of the BooNE experiment. It is searching for oscillations between muon and electron neutrinos. If an encouraging signal is seen at MiniBooNE, construction of BooNE will proceed.

MINOS (Main Injector Neutrino Oscillation Search): An experiment, that uses an accelerator at Fermilab to send a beam of neutrinos to the Soudan Mine to search for the oscillation of muon neutrinos.

Mixing Angles: A particle of a given flavor sometimes transforms into a similar particle of a different flavor. The parameters that quantify how likely this is to occur are known as mixing angles.

MRE (Major Research Equipment): The MRE account was created by the National Science Foundation for the funding and construction of large, cutting-edge research facilities.

Muon: A fundamental particle, identical to the electron, but approximately 200 times heavier. It is commonly found in cosmic rays.

Neutralino: A hypothetical electrically neutral supersymmetric particle. The superpartners of the photon, Z-boson, and Higgs Boson actually get blended together. These particles are called as neutralinos.

Neutrino: an uncharged, weakly interacting lepton, most commonly produced in nuclear reactions such as those in the sun. There are three known flavors of neutrino, corresponding to the three flavors of leptons. Recent experimental results indicate that all neutrinos have tiny masses.

Neutrino Oscillations: If neutrinos have mass, it is possible for them to convert from one flavor to another, and back again. This process is known as neutrino oscillation. If neutrinos oscillate from a flavor that is detectable into a flavor that is more difficult to detect, it may appear that the neutrinos have disappeared. This is a possible solution to the solar neutrino problem.

Neutron: One of the constituents of atoms. Along with the proton, the neutron is found in the nuclei (centers) of atoms. Neutrons have no electric charge, and are composed of two down quarks and an up quark.

NSAC: Nuclear Science Advisory Committee that reports to the DOE (Department of Energy) and the National Science Foundation.

NSF (National Science Foundation): One of the principal federal agencies supporting research in the physical sciences in the United States. Through its Division of Physics it provides approximately 30% of the support of the university High Energy Physics program.

PAC (Physics Advisory Committee): Each accelerator laboratory has a PAC to conduct rigorous reviews on proposed experiments and advise the laboratory director on the laboratory's program.

Parity: If a process is equally probable when the coordinates of space are reflected (as in a mirror), it is said to have parity symmetry.

Parton: The constituents of protons and neutrons are quarks, gluons and anti-quarks. They are collectively known as partons.

Parton Distribution Functions: These functions parameterize the composition (quarks, anti-quarks and gluons) of various hadrons. They describe the probability of finding a given constituent with a given momentum.

pb (picobarn): 10^{-12} barns. A unit used to measure cross-section. The inverse of this unit is used to measure integrated luminosity.

PEP II: An electron-positron collider, sited at SLAC, where the BaBar experiment is located.

PHENIX: An experiment based at Brookhaven National Laboratory (BNL).

Photon: The gauge boson that mediates the force of electromagnetism. The photon is the quantum of light.

Pierre Auger Observatory: A detector, currently under construction in Argentina that will be used to study the highest energy cosmic rays.

Pions: Pions are the lightest mesons. They are composed of up quarks, down quarks, and their anti-quark counterparts. Pions of charge +1, -1, and 0 are denoted pi^+, pi^-, and pi^0, respectively.

Positron: the anti-matter partner of the electron. It is identical in all respects to an electron, but it has positive charge.

p-p Collisions: Collisions between two protons.

Proton: One of the constituents of atoms. Along with the neutron, the proton can be found in the nuclei (centers) or atoms. Protons have a positive electric charge, and are composed of two up quarks and a down quark.

QCD (Quantum Chromodynamics): The theory of the strong interaction. It describes the exchange of gluons between quarks.

Quantum Dimensions: As yet undiscovered dimensions of space-time that convert force particles to matter particles and vice versa. Such dimensions are predicted in theories with supersymmetry or superstrings. The pair of particles that are converted into each other are called superpartners.

Quantum Gravity: At very small distances, the principles of quantum mechanics are necessary to accurately describe physical phenomena. Developing a theory that incorporates both the principles of quantum mechanics and gravity, a theory of "quantum gravity", has proven to be extremely difficult. String theory is the first real hope of providing such a theory.

Quantum Mechanics: In microscopic systems, particles such as electrons and protons behave like waves. Quantum mechanics replaces the more familiar Newtonian mechanics to describe such phenomena.

Quantum Numbers: The name given to the labels that describe various characteristics of elementary particles, atoms and molecules. Examples include the charge and spin of a particle.

Quark: One of the fundamental constituents of matter. They come in six flavors: up (u), down (d), strange (s), charm (c), bottom (b), and top (t). Only the up and down quarks are commonly found outside of accelerators. Protons and neutrons are composed of up and down quarks.

Rutherford Appleton Laboratory: Located in Oxfordshire in the United Kingdom, this laboratory has a broad-based research program that includes investigations in nuclear and particle physics.

SAGENAP: (Scientific Assessment Group for Experiments in Non-Accelerator Physics) A committee that reports to the Department of Energy and National Science Foundation.

Schrödinger's cat: Dead

Fermion: the superpartner of a fermion

SLAC (Stanford Linear Accelerator Center): Located in Menlo Park, California, this Department of Energy laboratory is the site of the BaBar experiment.

SLC (Stanford Linear Collider): This accelerator, located on the SLAC site, was completed in 1989 was used to study the Z-boson in detail.

SNAP (Supernova/Acceleration Probe): A proposed satellite-based experiment for finding and studying supernovae. The characteristics of such supernovae have proved useful in studying dark energy.

SNO (Sudbury Neutrino Observatory): This experiment, located 2 km beneath the surface in an active nickel mine in Sudbury, Ontario, is currently searching for neutrino oscillations in an attempt to disentangle the Solar Neutrino Problem.

Solar Neutrino Problem: The number of neutrinos observed to be coming from the sun is much less than the number predicted based on our understanding of the sun's inner workings. This discrepancy is known as the solar neutrino problem, and is one of the reasons to believe that there are neutrino oscillations.

Space-Time: See Special Relativity.

Special Relativity: Einstein discovered that time and space are interconnected. Height, width, length, and time make up the dimensions of space-time. The famous equation $E=mc2$ is a consequence of this theory.

Spin: A number that labels the intrinsic angular momentum of a particle, essentially how much the particle rotates around its axis.

This number can only take on discrete values. Particles with different spins will interact in different ways.

sQuark: The hypothetical spin-zero superpartner of the quark.

Standard Model: The standard model of particle physics is a theory of three of the four known fundamental interactions and the elementary particles that take part in these interactions. These particles are believed to make up all visible matter in the universe.

String Theory: A theory that seeks to incorporate a quantum theory of gravity into the Standard Model. In this theory, the fundamental constituents of matter are not particles, but strings. The particles that are observed are manifestations of the vibrations of fundamental strings.

Strong force: One of four known fundamental forces (the others are the weak force, electromagnetism and gravity). The strong force is felt only by quarks and gluons, and is responsible for binding quarks together to make hadrons. For example, two up quarks and a down quark are bound together to make a proton. The strong interaction is also responsible for holding protons and neutrons together in atomic nuclei.

SuperKamiokande: This experiment, located in the Kamioka mine in Japan, is the successor to the smaller Kamiokande experiment. SuperKamiokande provided compelling evidence for neutrino oscillations in neutrinos created by cosmic rays bombarding the Earth's atmosphere.

Supernova: When a star exhausts its nuclear fuel, it under goes a catastrophic collapse. The resulting explosion is known as a supernova. It often is brighter than an entire galaxy.

Superparticle: Supersymmetry predicts the existence of superpartners of the Standard Model particles. These new types of particles, the superparticles, would represent a new quantum dimension.

Superpartner: see superparticle.

Supersymmetry: A hypothetical symmetry relating particles of different spins. Under this symmetry, matter particles (spin one-half fermions) are related to force particles (spin-zero or spin-one bosons).

Superstrings: When supersymmetry is imposed on string theory, it becomes, superstring theory. The fundamental constituents of this theory are known as superstrings.

Symmetry: Physicists use symmetries to restrict possible theories of fundamental particles. As an analogy, consider the human face. If you were able to see only the right side of a person's face, you would still be able to guess what the other side of the face looks like, because of the symmetry that our bodies possess. The other side of the face cannot look like just anything, you know what it looks like-because of symmetry. Similarly, symmetries observed in nature limit what theories look like.

TESLA (TeV Energy Superconducting Linear Accelerator): A proposal for a next-generation linear collider made by an international collaboration based at DESY.

TeV (Tera-electron Volts): 10^{12} electron volts.

Tevatron: A 2 TeV proton on anti-proton collider that operates at Fermilab in Batavia, Illinois. The top quark was discovered using this accelerator.

Trilinear Coupling: The strength of an interaction involving three particles. In the case where all three particles are identical, it is known as the trilinear self-coupling.

Truther: Is one who rejects the official explanation provided for September 11, 2001. They include professional architects and engineers, scientists, and other scholars, firefighters, pilots, veterans, doctors, the victims' families and people from countless other walks of life. People who reject the official explanation live in a paradigm that sees the mainstream media as a tool of manipulation for the masses.

Architects and Engineers for 9-11 truth are the quintessential truthers.

Ultraviolet Catastrophe: The classical theory of light predicted that an ideal black body in thermal equilibrium would emit radiation with infinite power. This led to what became known as the Ultraviolet Catastrophe. See page 19 for more details.

Uncertainty Principle: In the world of quantum mechanics, there is an intrinsic uncertainty in studying the position and the momentum of a particle at the same time. This means studying physics at small distances, where an accurate determination of the position is needed, requires high momentum and hence high energy.

UNO (Underground Nucleon decay and Neutrino Observatory): A proposed Cherenkov detector experiment to search for proton decay, and neutrino oscillations.

VERITAS (Very Energetic Radiation Imaging Telescope Array System): A proposed experiment, based in the Arizona desert that will study gamma rays of slightly higher energy than the GLAST experiment.

Weak Force: This force is carried by heavy particles known as the W-boson and the Z-boson. The most common manifestation of this force is beta decay, in which a neutron in a nucleus is transformed into a proton, by emitting an electron and a neutrino.

Weak Neutral Current: A very weak interaction that is independent of the electric charge of a particle. Particles can exchange energy through this mechanism, but other characteristics of the particles remain unchanged. This force is mediated by the Z-boson.

WIMP: (Weakly Interacting Massive Particle): These as yet undiscovered particles are a leading candidate for dark matter.

WIPP (Waste Isolation Pilot Program): This underground cavern, located in Carlsbad, New Mexico, is a storage site for nuclear waste. It is also a proposed location for a National Underground Sciences Laboratory.

Z-boson: A gauge boson with no electric charge. It mediates the weak neutral current. It was studied in spectacular detailed by experiments at LEP and the SLC.

ZEUS: An experiment located at the DESY laboratory. ZEUS uses the electron-proton collider called HERA.

References

• Wikipedia: The free encyclopedia. (2004, July 22). FL: Wikimedia Foundation, Inc. Retrieved December 10, 2009, from: http://en.wikipedia.org/wiki/Black_body

• Wikipedia: The free encyclopedia. (2004, July 22). FL: Wikimedia Foundation, Inc. Retrieved November 10, 2009, from: http://en.wikipedia.org/wiki/Schrödinger_equation

• Stewart, I (2002). Does God Play Dice? The Mathematics of Chaos (pg 141). Wiley-Blackwell.

• Chaos Theory: A Brief Introduction. Retrieved December 10, 2009, from: http://imho.com/grae/chaos/chaos.html

• Wikipedia: The free encyclopedia. (2004, July 22). FL: Wikimedia Foundation, Inc. Retrieved 12/04/09, from: http://en.wikipedia.org/wiki/Standard_Model#Elementary_particles:_fer mions

• Standard Model (The Key to it all). Retrieved November 10, 2009, from: http://www.phy.bris.ac.uk/groups/particle/PUS/A-level/Standard_Model.htm

• Artist: Tiffany Ard. Retrieved March 10, 2010, from: http://www.babble.com/CS/blogs/strollerderby/2009/11/800px-Schrodingers_cat.png

• MacLean, G. (2009). Viennese Meow. Prima Storia

• Decoherence, the measurement problem, and interpretations of quantum mechanics. Maximilian Schlosshauer. June 28, 2005

• Wikipedia: The free encyclopedia. (2004, July 22). FL: Wikimedia Foundation, Inc. Retrieved December 18, 2009 from: http://en.wikipedia.org/wiki/Hilbert_spaces

• Universe is Finite, "Soccer Ball"-Shaped, Study Hints. Sean Markey. National Geographic News. October 8, 2003

• How has our universe evolved? Existing model confronted! Irani. Sep 8, 2008

• Thomas Levenson, September 2004, Discover Magazine

• His Lonely Path. Lee Smolin. Discover Magazine. Spring 2009.

• Relativity. Retrieved January 10, 2010, from: http://www.oglethorpe.edu/faculty/~m_rulison/Astronomy/Chap%2022/Relativity/relativity.htm

• Langevin, Paul (1911). "L'évolution de l'espace et du temps". Scientia 10: 31–54. Retrieved November 1, 2009, from: http://diglib.cib.unibo.it/diglib.php?inv=7&int_ptnum=10&term_ptnum=39&format=jpg

• Laue, Max von (1913). Das Relativitätsprinzip (2 ed.). Braunschweig: Vieweg.

• Morikawa, M. (2005) translation of A. Einstein "How I Constructed the Theory of Relativity," Translated by Masahiro Morikawa from the text recorded in Japanese by Jun Ishiwara, Association of Asia Pacific Physical Societies (AAPPS) Bulletin, Vol. 15, No. 2, pp. 17-19 (April 2005).

• WordPress.com. Retrieved January 10, 2010 from: http://thetechies.files.wordpress.com/2009/01/gravitational-lens-01.jpg

REFERENCES

• Bryant, Jeff; Pavlyk, Oleksandr. "Kepler's Second Law", Wolfram Demonstrations Project. Retrieved December 27, 2009

• Wikipedia: The free encyclopedia. (2004, July 22). FL: Wikimedia Foundation, Inc. Kepler's laws of planetary motion. Retrieved January 20, 2010, from:
http://en.wikipedia.org/wiki/Kepler's_laws_of_planetary_motion.

• Einstein, A. (1918) "Dialog über Einwände gegen die Relativitätstheorie", Die Naturwissenschaften 48, pp697-702, November 29, 1918 (English translation: dialog about objections against the theory of relativity)

• Jones, Preston; Wanex, L.F. (February 2006). "The clock paradox in a static homogeneous gravitational field". Foundations of Physics Letters 19 (1): 75–85. doi:10.1007/s10702-006-1850-3. Retrieved November 13, 2009, from: http://arxiv.org/abs/physics/0604025

• Robb, A. A. (1914): *A Theory of Space and Time. Cambridge*: Cambridge UP

• The Antinomy of Objects. Larvalsubjects. Retrieved January 20, 2010, from: http://larvalsubjects.wordpress.com/2009/01/24/the-antinomy-of-objects/

• "So what is String Theory Then?" Retrieved January 18, 2010, from: http://www.superstringtheory.com/basics/basic4.html

• Guth, A.H. (1998). The Inflationary Universe: Quest for a New Theory of Cosmic Origins. Vintage Books. ISBN 978-0099959502.

• Berkeley.edu. Retrieved December 04, 2009, from: http://cosmology.berkeley.edu/Education/ISTATPage/HighSchool/stellarE/ageu.jpg

• Schewe, P.; Stein, B. (2005). "An Ocean of Quarks". Physics News Update (American Institute of Physics). Retrieved January 10, 2010, from: http://www.aip.org/pnu/2005/split/728-1.html.

• Kolb and Turner (1988), Chapters 4, 6, 7

• Peacock (1999), chapter 9

• Poincare, in Capek, p. 323, and Robb, in Capek. pp. 369-386

• PREFACE TO THE ASTRONOMY AND COSMOLOGY SECTION. Nicholas M. Short, Sr. Retrieved January 3, 2010, from: http://rst.gsfc.nasa.gov/Sect20/preface.html

• Wikipedia: The free encyclopedia. (2004, July 22). FL: Wikimedia Foundation, Inc. Retrieved January 10, 2010, from: http://en.wikipedia.org/wiki/Length_contraction

• Creationwiki.org. Retrieved December 04, 2009, from: http://creationwiki.org/images/9/91/Srlc1.png

• (Hinshaw, G., et al. (2008). "Five-Year Wilkinson Microwave Anisotropy Probe (WMAP) Observations: Data Processing, Sky Maps, and Basic Results" (PDF). The Astrophysical Journal. Retrieved December 3, 2009, from: http://lambda.gsfc.nasa.gov/product/map/dr3/pub_papers/fiveyear/basic_results/wmap5basic.pdf

• Meyerson, in Capek, p. 355

• Newton, The Mathematical Principles of natural Philosophy, reprinted in Capek p. 209

• Race for 'God particle' heats up. James Morgan. Science reporter, BBC News, Chicago.

REFERENCES

• Einstein, A. (1905), "Ist die Trägheit eines Körpers von seinem Energieinhalt abhängig?", Annalen der Physik 18: 639–643, doi:10.1002/andp.19053231314

• Flores, F., E. N. Zalta, ed., The Equivalence of Mass and Energy, Stanford Encyclopedia of Philosophy. Retrieved February 17, 2010, from: http://plato.stanford.edu/entries/equivME

• Paul Allen Tipler, Ralph A. Llewellyn (2002). Modern Physics. W. H. Freeman and Company. pp. 87–88. ISBN 0-7167-4345-0.

• Einstein, in Capek, p. 361

• Simple Nature, by Benjamin Crowell. 1998-2009

• Young Two-Slit Experiment. Retrieved January 04, 2010, from: http://abyss.uoregon.edu/~js/21st_century_science/lectures/lec13.html

• Nuclear Fission. (unknown author) Retrieved January 4, 2010, from: http://www.kutl.kyushu-u.ac.jp/seminar/MicroWorld3_E/3Part3_E/3P33_E/nuclear_fission_E.htm

• NASA. Retrieved February 4, 2010, from: http://imagine.gsfc.nasa.gov/docs/science/know_l1/dark_matter.html

• NOAO Press Release 09-01: Elusive Binary Black Hole System. P. Marenfeld and NOAO/AURA/NSF

• Feynman, R. P.; Leighton, R. B.; and Sands, M. C h. 15 in The Feynman Lectures on Physics, Vol. 1. Redwood City, CA: Addison-Wesley, 1989.

• Krauss, L. M. The Physics of Star Trek. New York: Harper-Collins, 1995.

REFERENCES

• ScienceDaily (Nov. 16, 2006). Hubble Finds Evidence For Dark Energy In The Young Universe. Retrieved January 21, 2010, from: http://www.sciencedaily.com/releases/2006/11/061116132026.htm

• Time Machine. Peter Loader. Retrieved January 15, 2010, from: http://www.xaraxone.com/FeaturedArt/may04/html/08.htm

• Wikipedia: The free encyclopedia. (2004, July 22). FL: Wikimedia Foundation, Inc. Retrieved February 17, 2010, from: http://en.wikipedia.org/wiki/Wormhole

• Wikipedia: The free encyclopedia. (2004, July 22). FL: Wikimedia Foundation, Inc. Retrieved January 15, 2010 from: http://en.wikipedia.org/wiki/Image:Worm3.jpg

• Embedding Diagram. Soshichi Uchii. Retrieved January 17, 2010, from: http://www.bun.kyoto-u.ac.jp/~suchii/embed.diag.html

• 'Space-time cloak' could conceal events. Simon Hooper, CNN. Retrieved November 17, 2010, from: http://www.cnn.com/2010/TECH/innovation/11/16/space.time.cloak/index.html?hpt=T2

• Thorn, Kip S. (1994) Black Holes and Time Warps, Papermac, 1995.

• TIME TRAVEL: Strong evidence or major hoax??? UCalien. Retrived March 31, 2011 from: http://www.abovetopsecret.com/forum/thread554765/pg1

• The Circus. Charlie Chaplin. 1928 (silent film).

• Journal of Parapsychology, 60. 3-23. The American Institutes for Research Review of the Department of Defense's STAR GATE Program. March 1996.

• Double Dreamin'. Barbara and David P. Mikkelson. Retrieved February 15, 2010, from: http://www.snopes.com/luck/lottery/dreamwin.asp

REFERENCES

About the Author

I was going to begin this chapter by stating that this was the only part of the book that's 100% accurate. But that probably wouldn't be true. You see, there are very few things that I'm 100% certain of. I think Mark Twain put it best when he stated, *"the only certainties in life are death and taxes."* For all I know this entire world and universe could simply be a very clever illusion that we are all living in.

I originally wasn't even to going to include an *About the Author* section because I know it's rarely read. Most people just look at the picture and think terrible or perverted thoughts in their heads about the author. But nonetheless, for those of you who are compelled to know, I'll tell you a little about myself.

This book has been in the works for many years. While it has only taken a few dozen months to put in writing (actually closer to a few years), I've spent a very long time contemplating my thoughts. In fact, I remember reading scientific books and articles like the Special & General Theories of Relativity when I was about 12 years old. It was kind of like a "Relativity for Dummies" article, probably in a scientific magazine. I didn't understand much at the time, but I was fascinated about how these theories explained that time passed differently for different observers. Somehow these theories sparked a little curiosity in my brain and I was hooked on science. Of course, like most I was lured by the possibility of time travel.

I recently looked over my high school year book and saw that the career goal I listed for myself was to become a nuclear engineer. That partly explains why I didn't have many friends growing up, but I think my much conceded attitude at the time was the main reason. But I was ok not having many friends because one thing that I learned about friends is that unlike Chinese imports, quality is much more important than quantity. Thankfully, the few friends I do have I would trust with my life

Other than having a slightly warped brain and an odd personality my childhood was normal with very loving, somewhat compulsive parents who taught me, most importantly, how to be a good man. Growing up, I shared my home with a couple of fairly cruel siblings who enjoyed torturing their younger brothers, mostly through physiological warfare like shaving our heads and making us wear meat helmets on warm summer days. But that's what siblings do; and overall things turned out alright. And even today, on hot summer days, I still enjoy wearing a fresh meat helmet.

While in high school, my grades were a bit above average. I attribute this success entirely to the lack of fluoride in the water and the boringness of the Atari 2600, because I certainly never fully challenged myself academically. Instead, I managed to get into trouble on a near weekly or daily basis, mostly for schemes centered on making money and general pranks, humorous to only me and the severely disturbed.

After they let me out of high school (I think they just wanted to get rid of me), I entered the Navy and joined their nuclear propulsion program. Around this time, something in my brain clicked and I started doing pretty well in my job and later I did well in school. I think it just took me a little longer to grow up than it takes for most normal people. However, most people who know me would argue that I'm still not all the way there.

I went on to marry the love of my life, the beautiful Natasha. I later became the father of four great boys, each of whom is truly unique in his own personal way. Later on, I earned a Bachelor's of Science degree in Nuclear Engineering (pronounced noo-klee-er) from Rensselaer Polytechnic Institute and a MS in Applied Physics from the Naval Postgraduate School. I did all of this while holding my breath, or so it seemed. You see, time truly does pass differently for different observers. For me, the good times were quick and the bad times seemed to pass so slowly. I think this is one of God's many ways of reminding us that we're all still his *@?tch*. Pardon the pun, but how else can this be explained?

Speaking of God, I consider myself to be a spiritual man. And being such, it has always been my goal to understand how *He* thinks. I really enjoy reading scientific articles that attempt to disprove the existence of God. Sometime around 2010, Stephen Hawking, a leading and well known astrophysicist stated, "Because there is such a law as gravity, the universe can and will create itself from nothing. Spontaneous creation is the reason there is something rather than nothing, why the universe exists, why we exist." Hawking went on to further state that it is "not necessary to invoke God to light the blue touch paper and set the universe going."

"The religion that is afraid of science dishonors God and commits suicide." - Ralph Waldo Emerson

I find these types of ideas very disturbing. I'm very hopeful and confident in the existence of God. And it turns out that I'm in fairly good company. The following excerpts are taken from

Albert Einstein: *The Human Side*, Selected and Edited by Helen Dukas and Banesh Hoffman, Princeton University Press, 1979:

"A child in the sixth grade in a Sunday School in New York City, with the encouragement of her teacher, wrote to Einstein in Princeton on 19 January 1936 asking him whether scientists pray, and if so what they pray for. Einstein replied as follows on 24 January 1936:

I have tried to respond to your question as simply as I could. Here is my answer.

Scientific research is based on the idea that everything that takes place is determined by laws of nature, and therefore this holds for the actions of people. For this reason, a research scientist will hardly be inclined to believe that events could be influenced by a prayer, i.e. by a wish addressed to a supernatural Being.

However, it must be admitted that our actual knowledge of these laws is only imperfect and fragmentary, so that, actually, the belief in the existence of basic all-embracing laws in Nature also rests on a sort of faith. All the same this faith has been largely justified so far by the success of scientific research.

But, on the other hand, everyone who is seriously involved in the pursuit of science becomes convinced that a spirit is manifest in the laws of the Universe -- a spirit vastly superior to that of man, and one in the face of which we with our modest powers must feel humble. In this way the pursuit of science leads to a religious feeling of a special sort, which is indeed quite different from the religiosity of someone more naive."

It's reassuring to me to know that a man as brilliant as Einstein is able to admit that there are some things we still do not understand or know in science. It's worth mentioning that this letter was written a decade after the advent of Heisenberg's principle of indeterminacy and the probabilistic interpretation of quantum mechanics with its denial of strict determinism.

Lately it seems almost a requirement to attempt to discredit religion in articles, books and movies. I enjoy watching documentaries filmed by atheists who think they're smarter and better than everyone else. The simple fact is that they are not, in fact, quite the contrary. For whatever reason, these people think it's necessary to disprove God for their own self to exist. I think it's very important, especially nowadays, that we choose the people who we follow very wisely. There's little good we can learn from *most (but not all)* of the Hollywood stars, professional athletes, and the political elite.

I can never get enough of the stupid things that celebrities and politicians say. Here are just a few of my favorites. Remember when reading these quotes, that these were said by some of the most idolized people in our country:

"Smoking kills. If you're killed, you've lost a very important part of your life." *-Brooke Shields*

"I'm not going to have some reporters pawing through our papers. We are the president." *-Hillary Clinton*

"China is a big country, inhabited by many Chinese." *-Charles De Gaulle, former French President*

"Outside of the killings, Washington has one of the lowest crime rates in the country." -*Marion Barry, Washing, D.C. Mayor*

"I think that gay marriage is something that should be between a man and a woman. -*Arnold Schwartzanegger*

"I've got taste. It's inbred in me." -*David Hasselhoff*

"A zebra does not change its spots." -*Al Gore 1991 (and again in 1992)*

"Whenever I watch TV and see those poor starving kids all over the world, I can't help but cry. I mean I'd love to be skinny like that, but not with all those flies and death and stuff." -*Mariah Carey*

"I owe a lot to my parents, especially my mother and father." -*Greg Norman, Golfer*

"Those who survived the San Francisco earthquake said, 'Thank God, I'm still alive.' But, of course, those who died, their lives will never be the same again." -*California Senator Barbara Boxer*

I could go on and on with stupid quotes from celebrities; the point is that they may not be worthy of the praise we give them.

Many scientists too have some pretty crazy ideas and quotes. In fact, I may be one of those. It seems like there is an agenda for some scientists to attempt to prove that God does not exist. I'm not really sure what motivates this agenda. But if you don't agree with their viewpoint you are somehow less intelligent than they are.

But whether we consider books, new religions or films that discuss evolution, the big bang theory or the non-existence of God, they're all flawed for the same simple reason. You see, when discussing the big bang theory, one has to ask, "Where did the initial super-dense particle that *banged* come from?" Spontaneous creation is an answer that is far from sufficient to satisfy my curiosity. And if we can't answer that simple question, what the heck do we really know? I've come to accept the fact that I know so little but I remain anxious and optimistic of the possibility of learning so much more.

www.ingramcontent.com/pod-product-compliance
Lightning Source LLC
Chambersburg PA
CBHW020734180526
45163CB00001B/235